S0-BXE-547

INSTITUTE FOR RESEARCH INTO MENTAL RETARDATION

Study Group No. 2

CELLULAR ORGANELLES AND MEMBRANES IN MENTAL RETARDATION

IRMR STUDY GROUPS

1. Infantile Autism
2. Cellular Organelles and Membranes in Mental Retardation
3. The Brain in Unclassified Mental Retardation
4. The Application of Fundamental Research in the Behavioral Sciences to Practical Problems in Mental Retardation
5. Psychological Assessment of the Severely Subnormal

Cellular Organelles and Membranes in Mental Retardation

STUDY GROUP NO. 2

Study Group held at the Ciba Foundation, London under the auspices of the Institute for Research into Mental Retardation

EDITED BY

P. F. BENSON, PhD, MSc, MB, MRCP, DCH
Paediatric Research Unit,
Guy's Hospital Medical School,
University of London

Churchill Livingstone
Edinburgh and London 1971

First Published 1971

ISBN 0 7000 1519 1

Any correspondence relating to this volume should be directed to the publishers at 104 Gloucester Place, London W1H 4AE.

Printed in Great Britain by
The Whitefriars Press Ltd., London and Tonbridge

Preface

The Study Group brought together research workers with widely differing research orientations and with diverse types of technological expertise, but with a common interest, namely the study of biological phenomena at subcellular and molecular levels.

The advantages of a multidisciplinary approach to biological problems are evident from the discussions. There are numerous examples of how free exchange of ideas between participants engaged in different fields shed fresh light on some problems, and by posing many others suggested new ideas for research. Analogous situations were identified for widely differing biological phenomena, and experimental techniques familiar to some participants, were suggested for investigation of problems encountered by others.

The limited size of the Study Group encouraged thorough and informal discussion but restricted its immediate benefits to the few participants. It is hoped that this publication will help to overcome the latter drawback and prove useful to biochemists, biologists, clinicians, geneticists, neuropathologists and pharmacologists working in related fields.

P. F. BENSON

Acknowledgements

We are grateful to the Ciba Foundation for the Promotion of International Co-operation in Medical and Chemical Research for making this study group possible by offering their excellent facilities and accommodation at their London centre.

For financial assistance and support we are grateful to The Wellcome Trust and the Medical Research Council.

We are indebted to the staff of the Paediatric Research Unit, Guy's Hospital Medical School for their help in producing papers and to Professor Paul Polani for his advice and Chairmanship of the Study Group.

Contents

		Page
Preface		v
Acknowledgements		vi
Contents		vii

MEMBRANE CONDITIONED FACTORS

H. Bradford	Membrane potentials and metabolic performance in mammalian synaptosomes	1
H. McIlwain	Distribution and utilisation of amino acids in the brain	15
	Discussion	23

RENAL TUBULAR FUNCTION AND MENTAL SUBNORMALITY

K. Baerlocher, F. Mohyuddin and C. R. Scriver	Membrane transport of amino acids	29
D. N. Raine	Defects in renal tubular reabsorption of amino acids	43
J. B. Jepson	Hartnup disease	55
	Discussion	63

LYSOSOMAL FUNCTION AND DISEASE

J. T. Dingle	Lysosomes in nervous tissue	67
	Discussion	73
Helen Muir	Biochemistry of the mucopolysaccharidoses	75
	Discussion	93
R. J. Pollitt and	Aspartylglycosaminuria—a new inborn	

		Page
F. A. Jenner	error possibly due to a lysosomal enzyme defect	95
	Discussion	101
A. C. Allison	Types of lysosomal abnormality	107
	Discussion	115
N. B. Myant	Biochemistry of the gangliosides in relation to inborn disorders of ganglioside metabolism	117
	Discussion	131
D. N. Raine	Laboratory diagnosis of some sphingolipidoses	133
	Discussion	139

CHROMOSOMAL REDUNDANCY ASSOCIATED WITH MENTAL SUBNORMALITY

J. Stern	Biochemistry of Down's syndrome	143
	Discussion	157
P. F. Benson	RNA synthesis by Down's syndrome leucocytes	161
	Discussion	168
K. L. Yielding	The effects of chromosome redundancy on gene expression	173
	Discussion	182
H. Rees and R. N. Jones	Chromosome gain in higher plants	185
	Discussion	206
General Discussion		209
List of Participants		215
Index		217

Session I

Membrane Conditioned Factors

Membrane Potentials and Metabolic Performance in Mammalian Synaptosomes

H. F. BRADFORD

M.R.C. Metabolic Reactions Research Unit, Department of Biochemistry, Imperial College, London SW7

The cells of the nervous system often show a capacity to survive limited rupture or penetration of their membranes which might normally be expected to cause severe impairment of cell function through loss of soluble cytoplasmic cofactors and enzymes. Survival of such mechanical stress must be due either to the viscosity of the cytoplasm limiting the rate of losses through the damaged membrane, or due to a process of resealing of the membranes to prevent any losses occurring at all. Thus, pyramidal neurons in slices of cerebral cortex, and other excised brain regions maintained *in vitro,* are able to respond to invading synaptic potentials with action potential generation (Richards and McIlwain, 1967; Yamamoto and Kawai, 1967; Richards and Sercombe, 1968), and seem able to maintain high resting membrane potentials over periods of hours (Hillman and McIlwain, 1961; Gibson and McIlwain, 1965; Bradford and McIlwain, 1966) even though their axons were severed during preparation. Again, intracellular glass microelectrodes can penetrate and record for long periods from medium to large neurons even though the diameter of the shaft in the membrane may be in excess of 2 μ (Furshpan and Furukawa, 1962). Attempts to disaggregate the closely apposed and intertwined cellular units, which comprise the CNS, necessarily lead to cell damage. Thus, rupture and tearing away of the tenuous extensions of glial cells, and of the dendrites and axons of neurons, is inevitable. However, some preparations enriched in neuronal perikarya, or glial cells, show the capacity to oxidise and otherwise convert glucose at high rates with the formation of soluble products which remain occluded in the cytoplasm (Rose, 1967, 1968; Korey and Orchen, 1959).

Synaptosomes as Cell-like Entities

In view of the above considerations, it is not unexpected that nerve-ending processes, much smaller regions of the neuron, are able to survive liquid shear forces which disrupt neuronal and glial cell bodies, and form small, discrete, membrane-enclosed structures following rupture of their axon and of the membrane of the post-synaptic cell to which they are attached. These structures have been called 'synaptosomes' by Whittaker *et al.* (1964) and may be isolated by centrifugation procedures (Gray and Whittaker, 1962; De Robertis *et al.*, 1962).

Direct evidence that the membranes of these structures have resealed to form a continuous boundary comes from their osmotic properties, swelling and shrinking as they do in media of different tonicity (Marchbanks, 1967; Keen and White, 1970). When incubated, in the

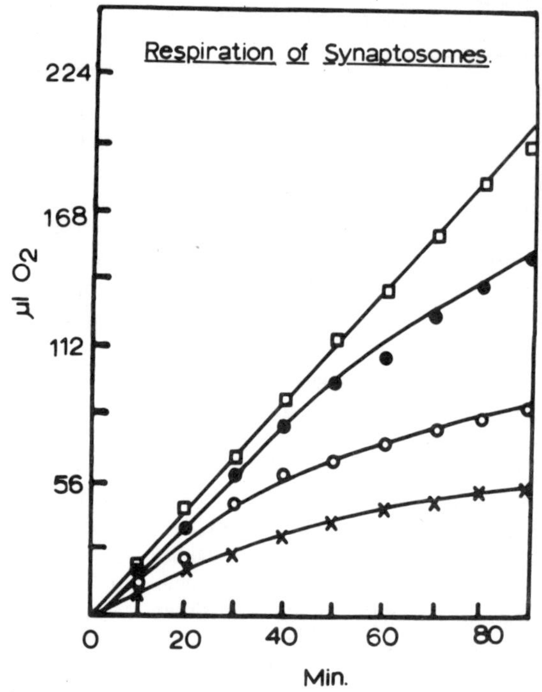

Fig. 1. Respiration of synaptosomes with glucose. Samples containing 10-15 mg of synaptosomal protein were incubated in 1.5 ml of Krebs-tris medium at 37° with substrate as follows: x–x, no substrate; ○ – ○, glucose 0.1 mM; ● – ●, glucose 1.0 mM; △ – △, glucose 5,mM (from Bradford, 1969).

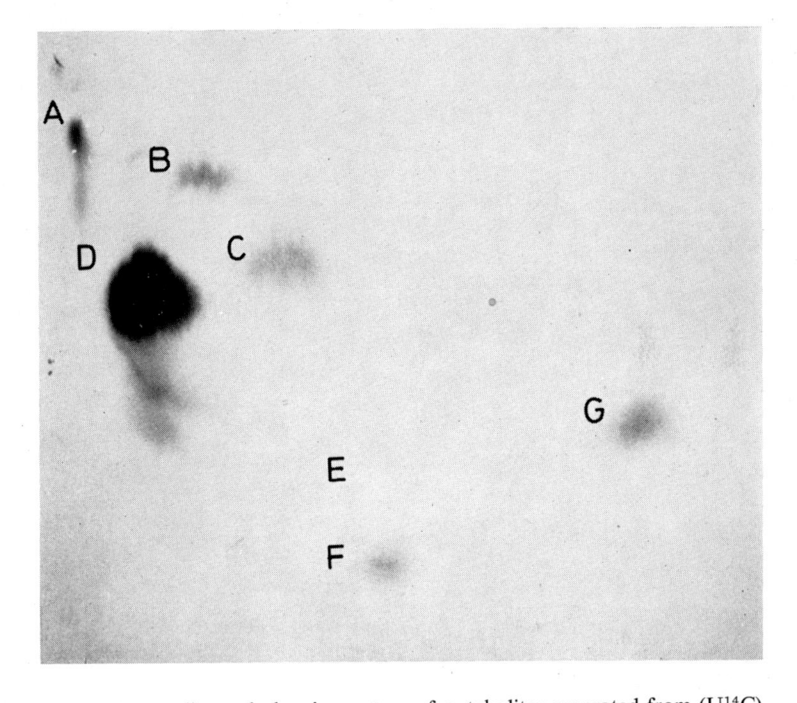

Fig. 2 Autoradiograph showing pattern of metabolites generated from ($U^{14}C$)-glucose during 1 hour's incubation in Krebs-phosphate medium containing 10 mM ($U^{14}C$)-glucose.
A, origin; B, aspartate; C, glutamate; D, glucose; E, Alanine; F, GABA; G, lactate.

[To face p. 2

presence of glucose, in media appropriate for whole cells, synaptosomes will in fact perform like whole cells, showing a high, linear respiration (Fig. 1), and producing lactate and amino acids (Fig. 2), generating ATP and phosphocreatine, and showing a capacity to accumulate potassium and extrude sodium against a concentration gradient (Bradford, 1967, 1969; Bradford and Thomas, 1969; Ling and Abdel-Latif, 1968). These properties suggest that synaptosomes provide a means of studying, *in vitro,* the metabolism of a specialised region of the neuron and its linkage to the biochemical events involved in neuro-transmission.

Membrane Potentials in Synaptosomes

The demonstration of a membrane potential existing across the synaptosomal membrane would provide a good index of its intactness, as well as providing evidence of the efficient functioning of membrane transport systems. The existence of these potentials can, however, only be established by inference since the small size of the synaptosome (0.5-0.2μ) makes direct measurement very difficult.

Evidence can be either in the form of data about the ion gradients across the membrane, or in terms of events and compositional changes occurring as the result of treatments expected to cause displacement of a membrane potential. The response of synaptosomes, incubated in nutrient salines, to oscillating electrical fields involves a great acceleration in oxygen uptake and lactate formation (Fig. 3) as well as a large decrease in potassium, and increase in sodium content (Fig. 4) (Bradford, 1970a, b). This is the complex of responses typically shown

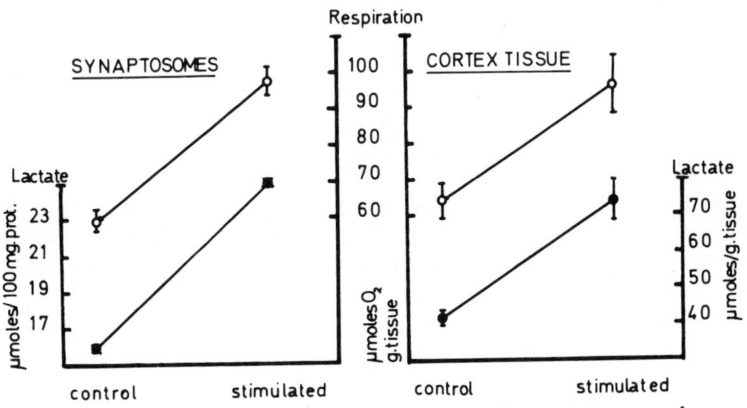

Fig. 3. Respiratory and glycolytic responses of synaptosomes and cortex tissue to electrical stimulation. Values are from ten preparations.

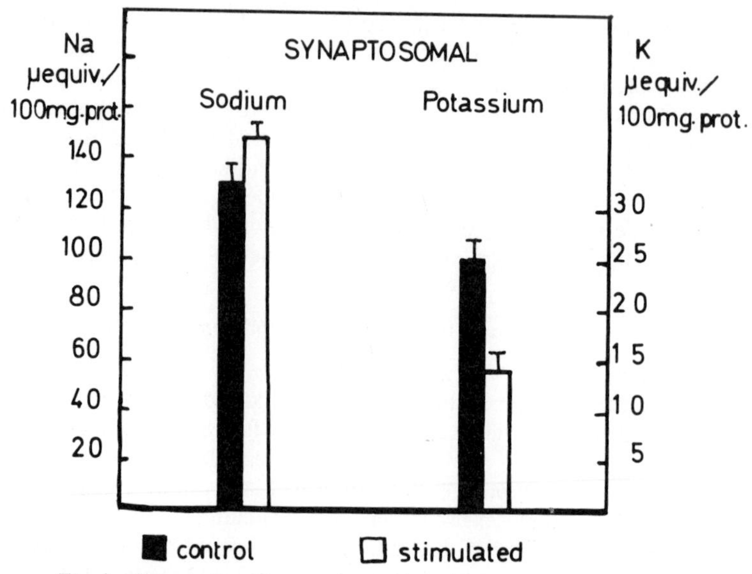

Fig. 4. Changes in sodium and potassium content of synaptosomes accompanying electrical stimulation. Values are mean of 10 preparations.

by neural cells undergoing depolarisation (Hillman *et al.,* 1963; McIlwain, 1963) and therefore provides evidence for the existence of a synaptosomal membrane potential.

Sodium and potassium contents of synaptosomes measured after incubation are shown in Table 1 together with the estimated concentrations of the ions in the non-inulin space. The latter values can be substituted in the modified Nernst equation (Hodgkin, 1958):

$$Em = \frac{RT}{F} \ln \frac{[\mathrm{K}]o + b[\mathrm{Na}]o}{[\mathrm{K}]i + b[\mathrm{Na}]i}$$

where o refers to extracellular concentrations, and i to intracellular concentrations, b is the apparent permeability of the membrane to Na relative to K (taken as 0.04, Bradford and McIwain, 1966), and Em is the membrane potential.

Such evaluation of the synaptosomal membrane potential gives a value of –27 mV, which is about half the potential expected in nerve-endings *in situ,* assuming that this would be close to the value measured in the perikaryon. However, if consideration is given to the likely influence on this estimate of the presence of amorphous membrane fragments in these preparations, which would tend to

TABLE 1.1

Sodium and Potassium Content of Synaptosomes

	Control	*Stimulated*
Sodium		
Content (μeq/100 mg protein)	145 ± 18 (10)	159 ± 21 (10)
Concentration (mM) in non-inulin space	15	31
Potassium		
Content (μeq/100 mg protein)	33 ± 2 (10)	18 ± 3 (10)
Concentration (mM) in non-inulin space	32	14

Values measured after 30 min preincubation in Krebs-phosphate, containing 10 mM glucose and 0.25% (w/v) inulin. Values are mean ± S.D. from number of experiments in parentheses (from Bradford and Thomas, 1970).

increase the non-inulin space, and consideration is also given to the fact that a large proportion of the synaptosomal volume is occupied by mitochondria, and synaptic vesicles (Gray and Whittaker, 1962; Jones and Bradford, 1970) which would reduce the space in the cytosol available for potassium, then −27 mV must be seen only as a minimal value for the membrane potential.

Apparent displacement of the synaptosomal membrane potential following application of electrical pulses can also be calculated from the data of Table 1 using the modified Nernst equation, and a change from −27 mV to −7 mV appears to occur when pulses are applied for 30 min.

Further basis for inferring the existence of a synaptosomal membrane potential comes from studies of calcium uptake of synaptosomes caused by progressively increasing medium potassium from 5 to 100 mM. The uptake rises steeply beyond 15 mM potassium which theoretically depolarises the membrane by 30 mV, and thereafter uptake follows the theoretical pattern of depolarisation (Blaustein and Wiesman, 1970).

Release of Amino Acids

In addition to the complex of responses described above, electrical stimulation and elevated medium potassium levels (6-55 mM) cause differential loss to the medium of certain amino acids, those showing loss in greatest proportion being aspartate, glutamate and GABA, all of which are strong candidates as transmitter substances in the cerebral cortex (Fig. 5, Fig. 6), and all of which are released *in vivo* from the cerebral cortex following stimulation or lesion of the mesencephalic reticular formation (Jasper and Koyama, 1969). However, although both potassium and amino acids are lost under these influences, the soluble enzyme lactate dehydrogenase shows only a small loss after 30 min and no detectable loss after 15 min of stimulation, though it is

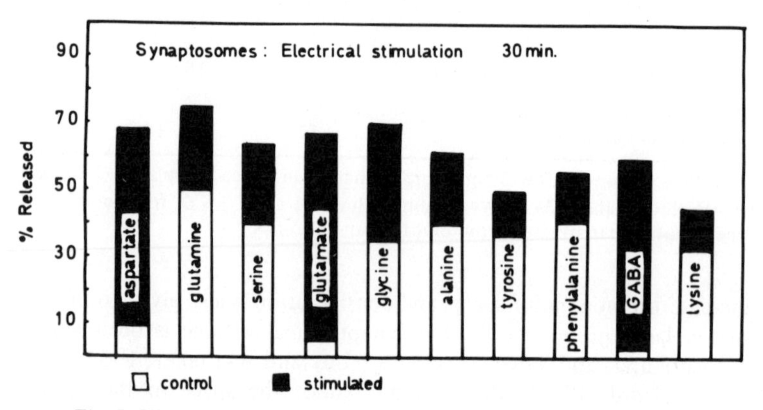

Fig. 5. Histogram showing proportions of total amino acids released to the medium in control and electrically stimulated synaptosomes (from Bradford, 1970b).

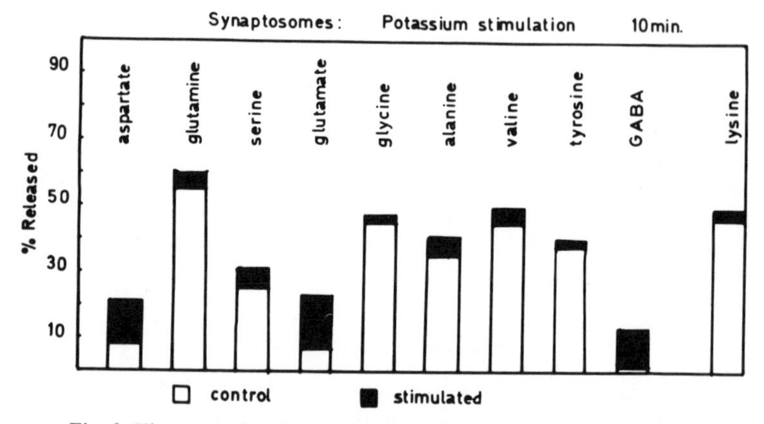

Fig. 6. Histogram showing proportions of total amino acids released to the medium by potassium stimulation. Values are mean of five preparations (from Bradford, 1970b).

TABLE 1.2

Lactate Dehydrogenase Content of Synaptosomes

	Lactate dehydrogenase activity (μ moles NADH oxidised/100 mg protein/hour x 10^2)	
	Tissue	*Medium*
Control	49 ± 6 (10)	17 ± 5 (10)
Stimulated		
15 min	48 ± 6 (10)	12 ± 3 (10)
30 min	42 ± 8 (10)	9 ± 2 (10)

The LDH activity was measured in medium and in water extracts of sedimented tissue after 60 min incubation in Krebs-phosphate-glucose, including periods of stimulation as indicated. Values are mean ± S.D. of number of experiments in paraentheses (from Bradford and Thomas, 1970).

readily released following osmotic rupture (Table 2). This indicates that electrical stimulation is unlikely to act by causing damage to the synaptosomal membrane rather through its depolarising influence, and certainly no obvious morphological damage occurs to the membrane following incubation or stimulation as judged from electron micrographs taken of positively fixed preparations (Jones and Bradford, 1970). Any heating effect must also be small since the temperature of the incubation fluid does not rise more than 0.3° C (Bradford, 1970a), and increasing the incubation temperature of synaptosomal suspensions from 37° to 42° C over a 5-min period accelerated both respiration and glycolysis—but was largely ineffective in releasing amino acids to the medium (a small loss of glutamate occurring inconsistently) (Table 3).

TABLE 1.3

Effect of Elevated Temperature on Synaptosomes

	37°C	*42°C*
Respiration (μmoles O_2/100 mg protein/hour)	61 ± 8 (9)	79 ± 11 (9)
Lactate accumulation (μmoles lactate/100 mg protein/hour)	24 ± 5 (9)	36 ± 7 (9)
Release of GABA, Glutamate and aspartate	none	none or small release of glutamate

After 30 min preincubation in Krebs-phosphate-glucose at 37°C the incubation temperature was raised to 42°C for 20 min. Values are mean ± S.D. for number of experiments in parentheses (from Bradford and Thomas, 1970).

Moreover, the kinetics of release show that the process is a relatively slow one, glutamate, aspartate and GABA being released during the whole of a period of 30 min stimulation (Fig. 7), which would not be expected in a preparation sustaining membrane damage of any magnitude.

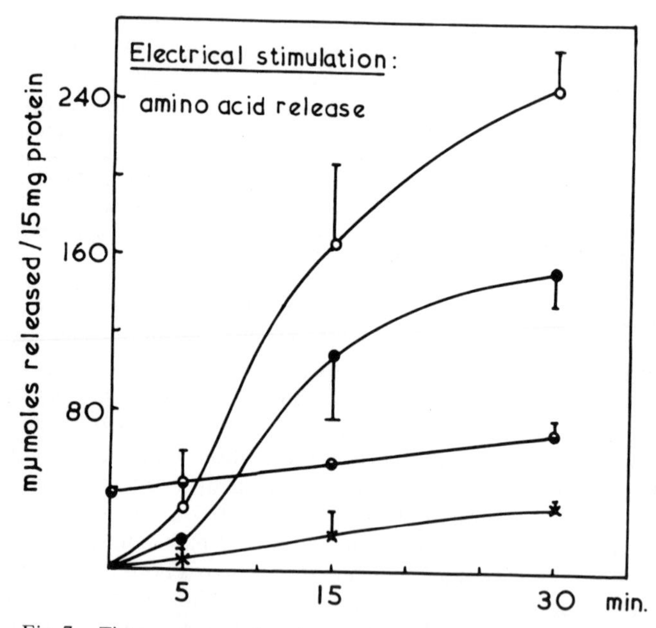

Fig. 7. Time course of release of some amino acids from synaptosomes caused by electrical stimulation. ○, glutamate; ●, aspartate; ○, alanine; x, GABA. Values are mean ± S.D. from six preparations.

Tetrodotoxin, chlorpromazine and cocaine have been tested for their effectiveness in preventing or attenuating these responses, and were found to be relatively ineffective, though all caused considerable inhibition of basal levels of glycolysis and respiration at concentrations which were without influence on these parameters in cortex tissue (Bradford, 1970a). Observations on the amphibian and mammalian neuromuscular junction, and the squid giant synapse, however, show that transmitter release mechanisms in the nerve terminal are remarkably resistant to tetrodotoxin (Katz and Miledi, 1965, 1969), which may be relevant in assessing the effects of drugs on synaptosomes under present experimental conditions.

Changes in the composition of the incubation medium which could be expected to condition the level of the membrane potential or the extent of depolarisation, such as using calcium-free media with or without complexing agents such as EGTA, profoundly influence both the basal metabolism and the response of synaptosomes (Table 4).

TABLE 1.4

Synaptosomal Respiration and Glycolysis in Calcium-free Medium

	Calcium Concentration (mM)	
	0.75	*0*
Respiration (μmoles O_2/100 mg protein/hour	58 ± 9 (10)	78 ± 12 (10)
Lactate Accumulation (μmoles lactate/100 mg protein)	26 ± 4 (10)	35 ± 7 (10)

Synaptosomes were incubated in Krebs-phosphate-glucose media containing 0.75 mM calcium, or without calcium but containing EGTA at 0.5 mM. Values are mean ± S.D. of number of experiments in parentheses (from Bradford and Thomas, 1970).

Thus, both respiration and glycolysis are greatly enhanced in calcium-free media but pulses are less effective in producing their acceleration, though the extent of the amino acid release remained high. This conditioning effect of calcium on the respiratory performance of synaptosomes is most likely to be through its influence on the sodium permeability of the membranes (Keesey *et al.*, 1965), and is an effect which is well established for cortical tissue *in vitro*.

Pulses and Amino Acid Turnover

The rate and extent to which glucose carbon is incorporated into aspartate, GABA, glutamate and alanine in synaptosomes is high (Bradford and Thomas, 1969), and electrical stimulation causes substantial enhancement of such incorporation into all of these amino acids except alanine (de Belleroche and Bradford, 1970).

Conclusions

The observations now accumulating from many laboratories together suggest that isolated nerve-endings retain much of their biochemical as well as their morphological complexity, and may, indeed, display many of the metabolic properties they possessed *in situ*, providing therefore, an excellent opportunity for study of the connexion between synaptic events and metabolism.

Summary

Synaptosomes from mammalian cerebral cortex show considerable biochemical versatility, performing in many ways like small whole cells when incubated in nutrient salines of appropriate composition. Evidence is presented to support the hypothesis that they are able to develop a membrane potential, an important property which would provide a good overall index of their capacity to accumulate certain ions and maintain their structure and metabolic integrity.

REFERENCES

Blaustein, M. P. and Wiesmann, W. P. (1970). Proceedings of the International Conference on Cholinergic Mechanisms in the CNS, Stockholm, 1970.
Bradford, H. F. (1967). Abstr. 1st Int. Neurochem. Congr., p. 30.
Bradford, H. F. (1969). *J. Neurochem.* **16,** 675.
Bradford, H. F. (1970a). *Brain Res.* **19,** 239.
Bradford, H. F. (1970b). Proceedings of the International Conference on Cholinergic Mechanisms in the CNS, Stockholm, 1970.
Bradford, H. F. and McIlwain, H. (1966). *J. Neurochem.* **13,** 1163.
Bradford, H. F. and Thomas, A. J. (1969). *J. Neurochem.* **16,** 1495.
Bradford, H. F. and Thomas, A. J. (1970). Submitted for publication.
de Belleroche, J. and Bradford, H. F. (1970). In preparation.
De Robertis, E., De Iraldi, A. P., De Lores Arnaiz, G. R. and Salgonicoff, L. (1962). *J. Neurochem.* **9,** 23.
Furshpan, E. J. and Furukawa (1962). *J. Neurophysiol.* **25,** 732.
Gibson, I. M. and McIlwain, H. (1965). *J. Physiol.* **176,** 261.
Gray, E. G. and Whittaker, V. P. (1962). *J. Anat.* **96,** 79.
Hillman, H. H. and McIlwain, H. (1961). *J. Physiol.* **157,** 263.
Hillman, H. H. Campbell, W. J. and McIlwain, H. (1963). *J. Neurochem.* **10,** 325.
Hodgkin, A. L. (1958). *Proc. roy. Soc. B.* **148,** 1.
Jasper, H. H. and Koyama, I. (1969). *Canad. J. Physiol. Pharmacol.* **47,** 889.
Jones, D. G. and Bradford, H. F. (1970). Submitted for publication.
Katz, B. and Miledi, R. (1965). *Nature, Lond.* **207,** 1097.
Katz, B. and Katz, B. and Miledi, R. (1969). *J. Physiol.* **203,** 689.
Keen, P. and White, T. D. (1970). *J. Neurochem.* **17,** 565.
Keesey, J. C., Wallgren, H. and McIlwain, H. (1965). *Biochem. J.* **95,** 289.
Korey, S. R. and Orchen, M. (1959). *J. Neurochem.* **3,** 277.
Ling, C. M. and Abdel-Latif, A. A. (1968). *J. Neurochem.* **15,** 721.
Marchbanks, R. M. (1967). *Biochem. J.* **104,** 148.
McIlwain, H. (1963). 'Chemical Exploration of the Brain', Elsevier, Amsterdam.
Richards, C. D. and McIlwain, H. (1967). *Nature, Lond.* **215,** 704.
Richards, C. D. and Sercombe, R. (1968). *J. Physiol.* **197,** 667.

Rose, S. P. R. (1967). *Biochem. J.* **102,** 33.
Rose, S. P. R. (1968). 'Applied Neurochemistry', Chapter 9 (eds A. N. Davison and J. Dobbing), Blackwells, Oxford.
Whittaker, V. P., Michaelson, I. A. and Kirkland, R. J. A. (1964). *Biochem. J.* **90,** 293.
Yamamoto, C. and Kawai, N. (1967). *Science* **154,** 341.

DISCUSSION

Jepson: When you say the cytoplasm is dense do you mean viscous?
Bradford: Yes.
Jepson: Is there any evidence that after electrical stimulation amino acids actually leave the synaptosomes or might they be synthesised *de novo* and released into the medium?
Bradford: In experiments lasting 30 min the evidence is that the total amount of amino acid (i.e. that in the medium plus that in the synaptosomes) remained the same. However, we have carried out some experiments for much shorter periods of stimulation in which the conditions are therefore more likely to be physiological, and we have some evidence that amino acids lost from the synaptosomes are replaced.
Scriver: Is the GABA pathway active in synaptosomes?
Bradford: I have not yet looked for this nor whether synaptosomes can use GABA as a substrate.
Scriver: Have you tried using amino-oxyacetic acid?
Bradford: No we have not. What we have done is to test the influence of drugs such as cocaine, tetrodotoxin and chlorpromazine. These substances, in equivalent concentrations, inhibit glycolysis and lactate formation in synaptosomes to a far greater extent than in brain cortex slice preparations. It is possible to select a concentration of these drugs which in synaptosomes will produce a 30% inhibition of glycolysis and a 10% inhibition of respiration but which has no detectable effect on cortex slice preparations.
Spector: What concentration of chlorpromazine did you use?
Bradford: 10 to 100 μM.
Spector: Are these concentrations within the range of those needed for pharmacological activity *in vivo*?
McIlwain: There are large differences in concentration of chlorpromazine in different areas in the brain; but after injection of pharmacologically active doses *in vivo* concentrations are in the order of 10 μM.

Scriver: You may be interested to hear that the complete GABA pathway is present in kidney and that its activity appears to be enhanced in ammoniogenesis. GABA itself is present in quite high concentrations in renal tissues. Its function in kidney is unknown but it seems unlikely that it acts as a neuro-transmitter. It seems more likely that it performs some function in a respiratory shunt; for example the pathway may offer an alternative route for the catabolism of glutamate and by the formation of succinate may be active in gluconeogenesis.

Bradford: This is very interesting. The metabolic role of GABA in kidney makes one wonder again whether the high concentration of GABA in the brain (which is probably greatly excessive for any requirement of transmission alone), may be related to metabolic functions similar to those you have postulated for kidney.

McIlwain: It is not clearly known why there is such a large quantity of GABA in cerebral tissue. A role as metabolic intermediate could presumably be met without large reserves of GABA. Its role apart from transmission is unclear. However, to retain such a large pool of GABA in the brain there must be a balance between the activities of enzymes which synthesise and degrade GABA, which is in favour of synthesis; the brain does not acquire the compound by assimilation after its synthesis elsewhere.

Jepson: Haven't differences been found in the glutamate decarboxylases of brain and kidney?

Scriver: Roberts and colleagues have claimed this. We have not yet substantiated their claim of enhancement of glial mitochondrial glutamic decarboxylase (type II) by amino-oxyacetic acid. We have found that the GABA shunt has been retained in the renal cortex and not in the papilla. I am wondering whether one can rationalise this (in evolutionary terms) by claiming any advantage to cell economy, if glutamate is metabolised via the GABA shunt rather than by oxidation to succinate via the Krebs' cycle.

Bradford: It might seem to be the reverse, since the utilisation of the GABA-shunt rather than the direct α-oxoglutarate pathway to succinate occurs at the expense of one ATP molecule, since there is no formation of succinyl-CoA in the former.

McIlwain: I think the capacities of the two systems differ greatly. Under many circumstances, by far the greater amount of glutamate is metabolised via the Krebs' cycle and only a small amount is used for the highly specialised production of GABA.

Yielding: It is interesting that the equilibrium of the first step of the GABA pathway, the glutamate dehydrogenase, is very much in favour of glutamate formation.

Have you carried out experiments with brain mitochondria and compared these with synaptosome preparations?

Bradford: Yes, both mitochondria and synaptosomes have been prepared from the same sample of brain and isolated in the same gradient. The mitochondrial preparations utilise mainly pyruvate for respiration and practically no glucose. They require ionic conditions similar to those present inside the cell, that is, the presence of high potassium and a low sodium concentration.
Spector: Can synaptosomes make protein?
Bradford: I cannot answer that with certainty. U-^{14}C-glucose is incorporated into substances present in the residue after water and ethanol extraction, which is probably protein. The specific radioactivity of this residue is about 10 times greater in experiments using synaptosomes than those using brain slices.
Spector: How long do synaptosome preparations remain active?
Bradford: Respiration continues for 2 hours at a constant rate, after which it begins to slow down.
Spector: Have you found that the conversion of radioactive glucose to amino acids is enhanced by electrical stimulation?
Bradford: Yes. When we stimulated synaptosomes sandwiched between two layers of nylon gauze we found that amino acids which were released to the medium by electrical stimulation were replaced by newly synthesised amino acids. Also, the specific radioactivities of glutamate and GABA were substantially increased by this stimulation.
McIwain: We should consider whether some of the amino acids released into the medium from stimulated synaptosomes may be derived in part from protein breakdown rather than by synthesis from glucose. Lysosomes may be needed for the degradation and disposal of protein which has arrived at nerve terminals. The amino acids released could then be returned to the extra cellular space.
Bradford: The presence of active lysosomes in cortex was demonstrated histochemically by Gordon *et al.* (1968, *Nature* **217**, 523) who showed activity of acid phosphatase in membrane-bound structures. It is certainly possible that degradation of protein by lysosomal enzymes might be a continuous physiological process rather than a feature of autolysis. However, in the experiments which I described there was a progressive increase in the specific radioactivity of amino acids in the medium. This indicated that they were synthesised *de novo* from radioactive glucose since if they had been derived from unlabelled protein they would not have been radioactive.
Allison: Have you actually measured lysosomal enzyme activity in synaptosomes?
Bradford: We have not but others have. During subcellular fractionation about half the lysosomal enzyme activity sediments with the mitochondrial fraction. The remainder is distributed throughout the gradient (e.g. Koenig *et al.*, 1964, *J. Neurochem.* **11**, 729).

Yielding: Could the differential effect on aspartate, glutamate and GABA be due to the presence of lysosomes in synaptosome preparations?

Bradford: No. This differential effect of electrical stimulation is likely to be a property of synaptosomes themselves, because of the occurrence of parallel changes in ion content and metabolism.

Dingle: First I should like to stress Professor McIlwain's point that the protein-degrading activities of lysosomes should be regarded as continuous physiological processes rather than occurring only during autolysis. I should also like to make the point that if one is interested in the function of lysosomal enzymes, such as proteases, one should test for these specifically. It is not justified to assume that proteases are present because one has demonstrated acid phosphatase activity.

Distribution and Utilisation of Amino Acids in the Brain in Relation to Cerebral Adaptation

HENRY McILWAIN

Department of Biochemistry, Institute of Psychiatry, (British Postgraduate Medical Federation, University of London) De Crespigny Park, London SE5

Amino acid disturbances are prominent among the known causes of mental retardation; this gives one reason for my describing here some of our current studies concerning the distribution and metabolism of amino acids in tissues of the brain. The second reason is that, at behavioural level, mental retardation has been described in terms of impairment of adaptation (Heber, 1961): a lack of appropriate response to environmental stimuli. When it is queried: how do organisms and organs in general adapt to environmental influences, the answer in biochemical terms is: by modulation of stages in the sequence from DNA to RNA categories and to protein effectors, which utilises nucleotides and amino acids. Until recently, there were few demonstrated examples of metabolic adaptation by these routes in the brain, but several are now available and are summarised in Table 2.1.

The types of response to environmental change, which are controlled in this way are quite varied. Some (1, 2, Table 2.1) concern chemical modifications: giving an amino acid, or the presence of a foreign substance. The third concerns response to visual signals and the fourth, to direct stimulation of the brain itself.

The fifth is perhaps that most relevant to the present theme, for it concerns experiments which are often interpreted as showing learning and retention. Such experiments, appraised for example by Glassman (1969) made it evident that changes at the level of DNA were not involved, but that it was by alteration at stages of RNA or protein metabolism in the brain that animals were enabled to change their performance.

TABLE 2.1

Cerebral Adaptive Processes and Evidence for their requiring Protein Synthesis

Process	*Metabolic Data*
1. Increased cerebral enzyme following glutamate administration	Increase in glutamate decarboxylase
2. Development of TOLERANCE to MORPHINE on continued administration	Associated with increased ^{32}P-incorporation to phospholipids; inhibited by puromycin or cycloheximide
3. Adjustment of daily and seasonal rhythms by ILLUMINATION: maintenance of certain gonadal functions	By pineal melatonin, the hydroxy-indole-O-methyl synthetase for its formation being induced by noradrenaline
4. Sustained increase in cortical CELL-FIRING by surface-positive polarisation	When established, persisted after cooling the cortex to 20°. Establishment blocked by cycloheximide or neomycin
5. Retention of certain induced BEHAVIOURAL responses: (a) for periods up to 30 min (b) for 30 min to 6 days	(a) Insensitive to inhibitors of RNA and protein synthesis (b) Blocked by acetoxycycloheximide and puromycin; actinomycin-D depressed and could act by diminished protein synthesis.

Data are mainly from the rat, for references and further details, see McIlwain, H. (1970a) Metabolic Adaptation in the Brain.

Cerebral Amino Acids and Protein Lability

These findings imply a lability and responsiveness in cerebral proteins and lead to consideration of the free amino acids of the brain from which the proteins are formed. In several of the amino acidurias associated with mental retardation the genetic defect has its quantitatively largest effect on amino acid metabolism in some part of the body other than the brain, and the brain is affected secondarily through change in amino acid concentrations in the blood. Moreover, although the initial abnormality may be in a single amino acid, the disturbance in amino acids of the brain can be more extensive, as is exemplified by the experimental phenylketonuria quoted in Table 2.2. Here the excess of phenylalanine induces a fall in the methionine and leucine of the brain. This is probably an example of competitive relationships among amino acids at membrane structures, which other contributors may appraise (see Blasberg and Lajtha, 1965; Neame, 1968). Included in Table 2.2 are experiments concerning two other conditions which can lead to mental defect: hypoxia and starvation in

TABLE 2.2

Examples of Conditions which Modify the Free Amino Acids of the Brain

Condition	*Changes in the free amino acid content of the brain*
PHENYLALANINE administration: 2 mg/g to infant rats (Agrawal *et al.*, 1970)	Entry of methionine and leucine to brain: −40 to −65% Phenylalanine of blood increased 12-20 fold
STARVATION of infant rats (Mourek *et al.*, 1970)	Alanine, aspartic acid, γ-amino-butyric acid, glutamic acid, glutamine: −10 to −63% in 10-14 days
HYPOXIA: 4.5% O_2 to dogs (Tews *et al.*, 1963)	Alanine, γ-aminobutyric acid, glutamic acid, leucine, tyrosine: +30 to +170% in 12 min Aspartic acid, methionine plus cystathionine: −30 to −40% in 12 min.

infancy. In these conditions also large changes take place in the free amino acids of the brain. The mental retardation of certain types of malnutrition has been associated with specific amino acid changes (see Scrimshaw and Gordon, 1968; Waisman, 1968). In hypoxia and in convulsions, the changes take place in a relatively short time (Tews *et al.*, 1963).

Administering isotopically-labelled amino acids shows that the cerebral free amino acids of normal animals undergo comparably rapid turnover, the half-lives of many being 20 to 50 min (Table 2.3). The turnover of amino acids of cerebral proteins are slower, but again allow scope for modification of protein make-up in relatively short periods. I would like to turn now to considering two amino acids in more detail from our current experiments (Fig. 1), which concerned the uptake of valine and leucine to isolated tissues from the brain, and their incorporation to protein. As free amino acids, these were distributed between tissue and fluid in favour of the tissue, by ratios of 4 to 8 which are comparable to their distribution between brain and cerebrospinal fluid *in vivo*. The time-course of their uptake showed equilibrium to be reached in about 30 min, and to proceed by a first-order reaction.

These experiments showed also a passage of amino acids from the tissues to the incubating fluids. This indeed would be expected for experiments both *in vivo* and *in vitro* have shown a flux of amino acids,

TABLE 2.3

Distribution and Turnover of Free Cerebral Amino Acids

Compound	*Content, μmoles/100 g or 100 ml*			*Turnover*[3]*: half-life, days*
	Brain	*Blood Plasma*[2]	*CSF*[2]	
Alanine	14	29	5	–
γ-Aminobutyrate	83	–	–	21
Arginine	6.9	16	2	–
Aspartic acid	245	2	–	
Glutamic acid	780	13	13	600
Glutamine	560	–	3	–
Glycine	55	29	23	–
Histidine	6.8	7	9	–
Leucine	8.9	12	2	14 – 34
Lysine	15	23	6	20 – 140
Phenylalanine	5.8	4	5	27 – 32
Serine	39	27	5	–
Threonine	29	13	2	–
Tyrosine	2.9	5	1	–
Valine	13	4	1	–

Sources of data: (1) Tews *et al.*, 1963, dog; (2) Bito *et al.*, 1966, dog; (3) see McIlwain and Bachelard, 1971, mainly rat.

outward as well as inward (Blasberg and Lajtha, 1965; Levi, Charayil and Lajtha, 1965; Neame, 1968). Also, the tissues are the source, by synthesis, of several amino acids, notably alanine, aspartic acid, glutamic acid, γ-aminobutyric acid and glutamine. This is not, however, the case with valine and leucine, which are dietarily essential. The feasible source of these amino acids is by protein breakdown. Data show (Table 2.4) that the output of valine and leucine from tissue to medium is considerable: after 1 hour, 85% of the free valine and leucine in the incubation system is in the fluid, while in the tissue the two amino acids have almost exactly the same concentration as in the brain *in vivo*.

The brain is known to be well provided with proteolytic enzymes (Marks and Lajtha, 1970) which have been separated to several fractions. In particular the cathepsin, acid-proteinase activity was found to be associated with dense particles with the properties of lysosomes (Marks and Lajtha, 1963; Koenig *et al.*, 1964). In fractions containing such particles the cathepsins and other lysosomal enzymes were activated by pre-incubation in presence of a non ionic detergent. This procedure solubilised the acid proteinases acting at pH 3.8, which could

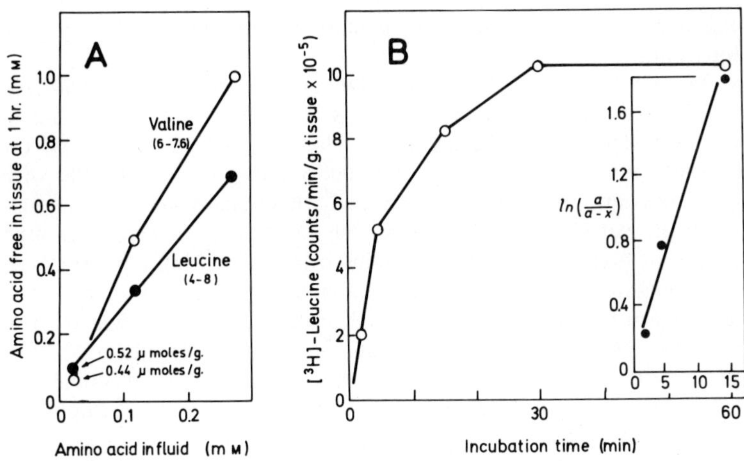

Fig. 1. Entry of amino acids to neocortical tissues from guinea pigs. A. Tissues were incubated for 1 hour in glucose-bicarbonate salines with the concentrations stated of valine and leucine, and later extracted for determination of the free intracellular levels of the two amino acids. The points nearest zero correspond to tissues with no added amino acids.

B. Entry of ^{3}H-labelled leucine; the inset gives a semilogarithmic plot indicating the initial entry to approach a first-order reaction (Jones and McIlwain, 1971).

TABLE 2.4

Free Valine and Leucine of Neocortical Tissue before and after Incubation

Tissue	*Amino acid, μmoles/g tissue*	
	Valine	*Leucine*
Fresh	0.08	0.07
After incubation		
in tissue	0.07 ± 0.01	0.10 ± 0.03
in fluid	0.44	0.52

Incubation was for 1 hour, of approx. 50 mg guinea pig tissue/5 ml glucose-bicarbonate saline (Jones and McIlwain, 1970).

also be prepared from acetone preparations of whole brain and were separable by electrophoresis to fractions with optimal activities at different peptide linkages, the major band corresponding to a cathepsin D. We calculated that the rate of amino acid formation during the experiments of Fig. 1 implied that the cerebral proteinases were then operating at rates corresponding to about 30% of the maximal rates found when they were optimally exhibited in tissue dispersions: a quite high value.

Protein turnover, by balanced and regulated processes of synthesis and breakdown is clearly important in all organs of the body but in neural systems an additional *spatial* factor is added because of the physical shape of neurons: often enormously elongated cells with cell-processes and the elaborate and functionally-important nerve-terminals some centimetres from the cell-body. There is organised supply of materials including proteins from the cell body by processes of axoplasmic transport. The preponderant movement of amino acids, also, is centrifugal: thus much of the protein components which arrive at the nerve-terminals presumably return extracellularly after proteolysis. It thus appears likely that the excretion of valine and leucine measured in the experiments of Table 2.4, reflects an important part of the tissue's normal amino acid economy. Note that the distribution in terms of amino acid *concentration* remains in favour of the tissue and that the quantity of amino acid appearing extracellularly may be connected with the experimental situation which places the tissue sample in contact with a much larger reservoir of extracellular fluid than exists *in vivo.* Output of several amino acids from small regions of the brain of monkeys, to slowly infused salines has been observed by DeFeudis *et al.* (1970). The quantitatively largest contribution, again, was by the group of amino acids synthesised in the brain. Under normal conditions the blood-brain barrier probably serves an important function in confining much of the flux of amino acids to the immediate interstitial fluid of the brain, as described elsewhere (see McIlwain, 1970b).

Stimulation and Localised, Persistent Changes in the Brain

Granted the flux of amino acids and the protein turnover which have been described, how are they directed to produce lasting and biologically-significant adaptation? This large subject, basic to mental retardation, merits raising here though its extent and complexity necessitate keeping to the systems concerned with the present theme. (In a wider view, morphogenetic mechanisms and processes of learning are involved.)

Thus (1) a demonstration that *protein synthesis* was modified by stimulation of the brain or of cerebral tissues, is relevant. Stimulation of the brain *in vivo* has been reported in some instances to increase and in others to decrease the incorporation of amino acids to protein (see Richter, 1966; Glassman, 1969). Stimulation of cerebral tissues *in vitro* has so far been associated only with decreased incorporation (Orrego and Lipmann, 1967; Lipmann, 1970). We confirm this (Jones and McIlwain, 1971) and have some additional conclusions. Namely, the incorporation appears to proceed from the intracellular pool of free amino acids. It can be diminished in association with the changes in intracellular Na, K and phosphocreatine caused by maximal electrical stimulation of the tissue, but under conditions of limited stimulation

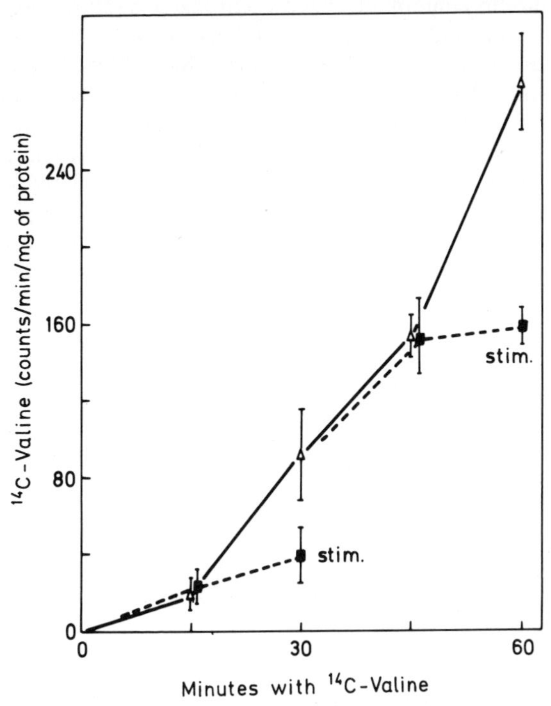

Fig. 2. The course of incorporation of ^{14}C-valine to the protein of guinea pig neocortex, incubated in glucose-bicarbonate salines. After 30 min preincubation, the valine was added; some tissues were electrically stimulated at 50 pulses/sec during the periods indicated with interrupted lines (Jones and McIlwain, 1971).

the change in amino acid incorporation is independent of these factors.

(2) In specifying the localisation of metabolic change brought about by stimulation, *transmitter release from nerve terminals* is of outstanding importance. The release of neurotransmitters and cognate substances presumably occurs from such sites. This release has also been reproduced in isolated tissues from the brain (Katz *et al.*, 1968; Srinivasan *et al.*, 1969; McIlwain and Snyder, 1970); several properties

show the release from the isolated tissue to be akin to that occurring *in vivo*.

(3) *Agents involved in neurotransmission* are normally pictured as causing primarily a brief cation movement, depolarization and cell firing. However, several, notably noradrenaline, serotonin and histamine have been found to bring about longer-lasting changes. They cause large increase in the content of cyclic adenosine-3′,5′-phosphate in tissues from the brain and this increase is augmented further by electrical stimulation of the tissues (Kakiuchi and Rall, 1968; Kakiuchi *et al.,* 1970). Cyclic AMP has proved an important agent in modifying enzyme activity; for example of cerebral phosphorylase and of protein kinases (Miyamoto, Kuo and Greengard, 1969; Weller and Rodnight, 1970), and in enzyme induction. It has the latter role in microorganisms, in reticulocytes and in the brain (Tao and Schweiger, 1970; Malkin and Lipmann, 1969; Klein *et al.,* 1970). A mechanism is thus provided for longer-lasting changes following stimulation. Granted that the neurotransmitters concerned may evoke enzyme induction as well as cell-firing, additional purpose may be seen in the special cytological and enzymic arrangements for localising and terminating their action (see McIlwain, 1970). It is by the sequence outlined that one sees sensory input from the environment modifying the normal brain; and in this sequence, therefore, that understanding of the retarded development in mental defect may be sought.

ACKNOWLEDGEMENT

I am grateful for support from the Research Fund of the Bethlem Royal Hospital and the Maudsley Hospital and from the Science Research Council, during these investigations.

REFERENCES

Agrawal, H. C., Bone, A. H. and Davison, A. N. (1970). *Biochem. J.* **117,** 325.
Bito, L., Davson, H., Levin, E., Murray, M. and Snider, N. (1966). *J. Neurochem.* **13,** 1057.
Blasberg, R. and Lajtha, A. (1965). *Arch. Biochem. Biophys.* **112,** 361.
DeFeudis, F. V., Delgado, J. M. R. and Roth, R. H. (1970). *Brain Res.* **18,** 15.
Glassman, E. (1969). *A. Rev. Biochem.* **38,** 605.
Heber, R. (1961). 'Manual on Termination and Classification in Mental Retardation'. Amer. Assoc. Ment. Defic.
Jones, D. and McIlwain, H. (1971). *J. Neurochem.* **18,** 41.
Kakiuchi, S. and Rall, T. W. (1968). *Mol. Pharmacol.* **4,** 367, 379.
Kakiuchi, S., Rall, T. W. and McIlwain, H. (1970). *J. Neurochem.* **16,** 485.
Katz, R. I., Chase, T. and Kopin, I. J. (1968). *Science* **162,** 466.
Klein, D. C., Berg, G. R. and Weller, J. (1970). *Science* **168,** 979.

Koenig, H., Gaines, D., McDonald, T., Gray, R. and Scott, J. (1964). *J. Neurochem.* **11,** 729.
Levi, G., Cheravil, A. and Lajtha, A. (1965). *J. Neurochem.* **12,** 757.
Lipmann, F. (1970). 'Protein Metabolism of the Nervous System', p. 305. New York: Plenum Press.
Malkin, M. and Lipmann, F. (1969). *Proc. Nat. Acad. Sci. U.S.* **64,** 973.
Marks, N. and Lajtha, A. (1963). *Biochem. J.* **89,** 438.
Marks, N. and Lajtha, A. (1970). 'Protein Metabolism of the Nervous System', p. 39. New York: Plenum Press.
McIlwain, H. (1970a). *Nature, Lond.* **226,** 803.
McIlwain, H. (1970b). Wates Symposium on the Blood-Brain Barrier. Oxford: University Lab. of Physiol.
McIlwain, H. and Bachelard, H. S. (1971). 'Biochemistry and the Central Nervous System'. 4th Edition. London: Churchill.
McIlwain, H. and Snyder, S. H. (1970). *J. Neurochem.* **17,** 521.
Miyamoto, E., Kuo, J. F. and Greengard, P. (1969). *Science* **165,** 63.
Mourek, J., Agrawal, H. C., Davis, J. M. and Himwich, W. A. (1970). *Brain Res.* **19,** 229.
Neame, K. D. (1968). *Progr. Brain Res.* **29,** 185.
Orrego, F. and Lipmann, F. (1967). *J. biol. Chem.* **242,** 665.
Richter, D. (1966). In 'Aspects of Learning and Memory' (ed. D. Richter), p. 73. London: Heinemann.
Scrimshaw, N. S. and Gordon, J. E. (1968). 'Malnutrition, Learning and Behaviour'. Cambridge, Mass. MIT Press.
Srinivasan, V., Neal, M. J. and Mitchell, J. F. (1969). *J. Neurochem.* **16,** 1235.
Tao, M. and Schweiger, M. (1970). *J. Bact.* **102,** 138.
Tews, J. K., Carter, S. H., Roa, P. D. and Stone, W. E. (1963). *J. Neurochem.* **10,** 641.
Weller, M. and Rodnight, R. (1970). *Nature* **225,** 187.
Waisman, H. A. (1968). *Proc. Congr. Internat. Assoc. Sci. Study Ment. Defic.* **1,** 514.

DISCUSSION

Robinson: Is there any explanation for the remarkably long half-life of glutamate?

McIlwain: This relates to the very large amount of glutamate present in the brain, that is 780 μ mols/100 g as compared with the concentration of say valine which is about 13 μ mols/100 g, roughly one sixtieth. If the turnover on a molar basis were the same for glutamate and valine the half-life for glutamate would be very much greater than for valine.

Jepson: Do you include glutamine in your figure for glutamate?

McIlwain: No, but there is also quite a high concentration of glutamine in the brain.

Yielding: When one considers catabolism of glutamate, the fate of the amino group is probably more important than that of its carbon skeleton since the former contributes to the synthesis of the other amino acids.

Allison: I imagine that you used labelled valine for experiments on protein synthesis because its concentration in the brain is normally low. I noted that if one extrapolates the curve backwards, it seems that synthesis of one protein molecule takes about 15 min which seems quite reasonable. Does the rate of protein synthesis (and presumably breakdown) occur more quickly in certain parts of the brain than in others?

McIlwain: Yes, for example, the S-100 protein is of relatively rapid turnover, and that of myelin is slow. The average half-life of cerebral proteins is about a week but some proteins have a half-life of 5-10 hours, while others are of some weeks.

Jepson: The average value need not necessarily be important. Some essential but small fractions of protein may have a very small half-life.

Allison: Presumably some proteins are much more stable. After all memory which can last for years must have some structural basis.

Myant: You spoke of cyclic 3′,5′-AMP (3′,5′ AMP) as being concerned with protein synthesis. I have always thought of it as being concerned with activation.

McIlwain: There are some instances in which neurotransmitters appear to increase the rate of specific enzyme protein synthesis by release of 3′,5′-AMP. For example, in pineal cultures noradrenalin can be shown to increase production of catechol-O-methyl-transferase which catalyses the rate-limiting step of the conversion of 5-OH tryptamine into melatonin, and there is evidence that this is mediated by 3′,5′-AMP (Axelrod *et al.*, 1969, *Proc. U.S. Nat. Acad. Sci.* **62**, 544; Klein *et al.*, 1970, *Science* **168**, 979). Other examples are enzyme induction in mammalian cells by steroids, noradrenalin and serotonin; but not by acetylcholine which does not appear to act via 3′,5′ AMP in the nervous system. Enzyme induction mediated by 3′,5′AMP has been reported also in microorganisms.

Spector: In addition to this role in enzyme synthesis which Professor McIlwain has just described, 3′,5′-AMP appears to be active in the initiation of mRNA synthesis in bacteria during the process of enzyme induction (Paston and Perman, 1970, *Science* **169**, 339).

Jepson: Do you believe that a neurotransmitter is directly concerned with the conversion of ATP to 3′,5′ AMP or does it act by stimulation of cyclase?

McIlwain: I think it acts through stimulation of cyclase which catalyses the conversion of ATP to 3′,5′ AMP.

Bradford: Do you have an estimate of the amount of leakage of the cell content that might occur from sliced cells of brain-slice preparations as compared with leakage from intact cells?

McIlwain: When calcium is present in the medium there does not appear to be any great initial leakage from sliced tissues. It is possible that damaged cells may be rapidly re-sealed–a phenomenon which has been demonstrated both in central nervous tissue and in cardiac muscle fibres in the presence of calcium.

Allison: Yes, and cardiac muscle fibres can then contract perfectly normally. This question of re-sealing brings one back to Dr. Bradford's point of whether the synaptosomes really are re-sealed or not. There is plenty of evidence that phospholipid structures tend to reseal rapidly.

Bradford: One can visualise phospholipid resealing but it is more difficult to see how protein structures can be repaired so easily.

Dingle: One can visualise that initially resealing is by phospholipid and later on by protein.

Bradford: It is my view that synaptosomes probably are sealed, but at present this is not proven. The property of resealing of whole cells may be investigated by measurement of membrane potentials. However, results may vary widely according to the size of electrodes used. I think that microelectrode terminals should not be greater than about 2 μ in diameter, and even then, in practice, they may penetrate to a depth which leaves the diameter of the barrel passing through the membrane at more like 4 μ. Electrodes with tips larger than 2 μ will cause a fall in membrane potential.

Allison: The situations in whole neurons are not strictly comparable to those in synaptosome preparations since in the former the structures are being held together.

Jepson: One should also consider whether in disrupted cells molecules can enter before re-sealing has occurred as has been demonstrated for red cells.

McIlwain: Artefacts may also be caused by contamination with salts derived from within the electrode and introduced through the damaged membrane.

Bradford: An estimate of lasting damage inflicted on cell membranes by electrode penetration can be made by measurement of membrane potentials on a second penetration. It is generally found that the membrane potential at second penetration is often at a lower level than it was at the time of the first. I think the problem may be that conduction of current occurs from inside to outside of the cell along the polar surface of the electrode itself, and thus these electrode measurements are not good indices of membrane re-sealing.

Yielding: Have you studied the uptake of amino acids by synaptosomes?

Bradford: I have not done this myself but it has been studied by Roberts' group in California (Varon *et al.*, 1965, *Biochem. Pharm.* **19**, 1755), who found that GABA is actively taken up in the presence of sodium. Uptake of glutamate is more difficult to assess, since it behaves differently when it is on the inside to when it is outside. In the latter case it causes depolarisation and a reduction in tissue ATP concentration. I have studied the utilisation of ^{14}C-glutamate and found it is very rapidly metabolised by synaptosomes into aspartate and GABA, and this is accompanied by oxygen uptake. In fact, GABA formation is about 10 times more rapid from glutamate than from glucose.

Dingle: What information is there about membrane thickening?

Bradford: In the past few years there has been an enormous growth in the observations of structures which can be seen in a synapse stained for electron microscopy, and also in the precise form of well-known structures. Thus, the pre- and post-synaptic cells are separated by a synaptic cleft 160-200 Å wide and filled with a dense granular material. Similar dense material also accumulates on the intracellular side of each membrane. On the pre-synaptic side this forms the dense projections, the pre-synaptic network, and the pre-synaptic grid and on the post-synaptic side it forms the post-synaptic thickening. Peters (1969, *Z. Zellforsch.* **100**, 487) has shown that this material is not an integral part of the boundary membranes which appear to exist independently of it. Also, it seems that the post-synaptic thickening in a dendrite spine is commonly an annulus or perforated disc which gives the appearance of a double or multiple thickening in vertical section. Vesicles usually accumulate on the pre-synaptic side only over the area of the thickening, and presumably this indicates a strong functional connection between these two types of structure.

McIlwain: Are these structures associated with microtubules?

Bradford: There is usually a mass of very thin filamentous whispy material in the dendrite spine which receives the axon terminal of the pre-synaptic cell, and motoneurone synapses commonly have a ring of microtubule from the axon on the pre-synaptic side.

Allison: This is extraordinarily like the centriole and the basis of the mitotic spindle. We are very interested in this because we think that general anaesthetics may work by causing depolarisation of microtubules. There are indications that this occurs at certain types of synapses but not at others.

Polani: Is there any evidence that synaptosomes form a single population?

Bradford: Quite the reverse. The evidence is overwhelming that synaptosomes from cerebral cortex consist of a very mixed population of structures.

Polani: Can one separate the various types of synaptosomes into preparations of more homogenous particles?

Bradford: There have been a number of attempts at this, e.g. de

Robertis *et al.* (1962, *J. Neurochem.* **9**, 23) used a second density layer in his gradients and were able to separate two populations of synaptosomes.The lighter population was rich in acetylcholine esterase and choline acetylase. The denser population was associated with glutamate decarboxylase and glutamate dehydrogenase activities.

Polani: Is there any evidence on what happens to synaptosomes during development?

Bradford: Abdel-Latif (1967, *J. Neurochem.* **14**, 1133) described some developmental changes of sodium-dependent ATP activity and also in ^{32}P incorporation of isolated nerve-endings (Abdel-Latif, 1968, *Brain Res.* **10**, 307), I cannot quote any studies which have followed the morphological development of nerve-endings or their derived synaptosomes.

McIlwain: It has been claimed that the number of synaptic junctions in the visual system can be diminished by blinding animals.

Raine: Amino acids have very high potential for altering dielectric constants. Has this property been considered as a possible mechanism in relation to their effect on membrane potentials?

McIlwain: I think the effect which glutamate has in bringing about at 0.2 mM a large influx of sodium ions is far greater than could be effected by changes in dielectric constants.

Stern: I was struck by the considerable uptake of methionine and leucine in the presence of phenylalanine and wonder whether these observations may be related to the abnormal biochemical environment in phenylketonuria. There is an increasing amount of evidence which suggests that the cerebral abnormality in phenylketonuria might be due to interference with transport of amino acids in the presence of high concentrations of phenylalanine. I have also wondered why the clinical prognosis is so severe in disorders in which there is accumulation of leucine. It would be interesting to know whether high concentrations of leucine could interfere with transport of other amino acids.

Scriver: We were intrigued with the theory on the nature of anaesthetic substances put forward by Pauling. We wondered whether, in the aminoacidopathies, cerebral damage is produced by perturbation, first of amino acid balance and second of the structure of cellular membranes and brain tissues. However, no evidence for this was found by Neal at McGill.

Allison: Pauling's theory of anaesthesia still remains to be proved.

Pollitt: In non-ketotic hyperglycinaemia the administration of glycine can produce very toxic effects. Can anyone explain this?

Komrower: Some years ago we observed that glycine administration in this condition produced severe hypoglycaemia. This was reminiscent of leucine induced hypoglycaemia.

Jepson: Do you get hypoglycaemia with maple syrup urine disease?

Komrower: Yes. Quite severe hypoglycaemia.
McIlwain: As the convulsive effects of strychnine may be related to interference with glycine metabolism, I should like to know what were the clinical features of the disturbance following glycine administration in the patient with non-ketotic hyperglycinaemia?
Komrower: The main features were that after a glycine load the infant became semi-comatose and there was some generalised twitching. When his level of consciousness recovered he was ataxic and this persisted for some time. In addition there were episodes of ataxia which occurred without feeding with glycine, usually in relation to an infection, for example, measles.
Bradford: In high concentrations systemic glycine may be acting as an inhibitory transmitter in the spinal cord and the medulla. A comparable situation is found with glutamate which in very high oral doses causes arousal from hypoglycaemic coma (Petersen *et al.*, 1955, *Am. J. Physiol.* **181**, 519). On the other hand when injected into the cerebral ventricles in small amounts it causes paroxysm (Weichert 1968, *J. Neurochem.* **15**, 1625).
Allison: The effect of high concentration of amino acids produced by intravenous infusion could be studied in experimental animals. It would be interesting to study the metabolism of labelled amino acids *in vivo* and to compare this with their metabolism *in vitro* in the type of systems described by Professor McIlwain and Dr. Bradford.
Raine: Dr. Scriver has infused glycine into humans with imino-acid-glycinuria.
Scriver: We have studied the effect on kidney function but not on brain. We loaded a patient with proline, and although the plasma levels rose 10-fold we could not detect any increase in his C.S.F. proline concentration. It would have been very interesting to know whether brain metabolism was affected. Some time ago I tried to correlate the concentration of amino acids in the C.S.F. with that in the tissues in which they were metabolised. The results indicated that the concentration of a particular amino acid in the C.S.F. depended in part on whether it was metabolised by the brain rather than on its blood concentration.
Benson: This has also been observed in argininosuccinicaciduria where levels of argininosuccinic acid (A.S.A.) were higher in the C.S.F. than in the plasma (Allan *et al.*, 1958, *Lancet,* **1**, 182). This was interpreted as indicating that A.S.A. is normally metabolised by the brain and that in this disorder the enzyme defect led to its accumulation in the C.S.F.

Session II

Renal Tubular Function and Mental Subnormality

Membrane Transport of Amino Acids; Use of Mutation and Ontogeny as Probes of Specificity

KURT BAERLOCHER,
FAZL MOHYUDDIN
and
CHARLES R. SCRIVER

The deBelle Laboratory for Biochemical Genetics
The McGill University-Montreal Children's Hospital
Research Institute
2300 Tupper Street
Montreal 108, Quebec, Canada

Two important mechanisms control the intracellular composition of free metabolites: enzymes direct the conversion of one molecular form to another; and membrane transport processes control fluxes of metabolites into and out of cellular pools.

A membrane which controls permeation of amino acids can play an important role in cellular economy, protein synthesis, growth and development. And yet our understanding of the cell and particularly of its plasma membrane has grown slowly. It is only 130 years since Nageli, the German botanist, defined the existence of the plasma membrane (Scott, 1891), thus completing the basic concepts of cell theory. Although Nageli could test for the presence of the plasma membrane by simple osmotic experiments he could not see it.

Collander and Barlund (1933) noted that the plasma membrane was a peculiar permeability barrier; many non-electrolytes penetrated it at rates in accordance with their lipid solubility, and yet many polar solutes entered too freely if the membrane were merely a lipid barrier to the permeation of hydrophylic molecules. These and other observations led Danielli and Davson (1935) to propose their paucimolecular hypothesis of membrane structure.

Today we can 'see' a membrane under the electron microscope, or we think we can, but it is still only remotely probable that the familiar 'railroad-track' artefact of modern microscopic techniques (Robertson,

1964) is anything like the living plasma membrane. Newer methods of physical and chemical analysis (Korn, 1969) have eroded the familiar Davson-Danielli hypothesis, making it difficult to reconcile the earlier bimolecular lamellar model with present data. Chemical, kinetic and genetic probes (Christensen, 1967; Christensen, 1969; Scriver and Hechtman, 1970) reveal a membrane of apparently non-rigid structure containing transport proteins (Pardee, 1968) within a lipid matrix, which can mediate the flux of organic solutes such as amino acids. Transport of these molecules is achieved by a conjugate driving force (Schultz, 1969), coupled usually to the flow of other solutes or liquid, and to metabolic energy, to achieve polarity of the flux and to maintain transport against an electro-chemical gradient. The conjugate driving force operates with great specificity towards its substrate (Berlin, 1970) and it observes classical Michaelis-Menten kinetics; moreover it is under complex genetic control (Rosenberg, 1969; Scriver, 1969; Scriver and Hechtman, 1970). The plasma membrane is thus revealed as a mosaic of reactive sites which regulate the coupled fluxes of solutes. This can be illustrated well by a resumé of some of our work on the membrane systems for iminoglycine transport in kidney.

The Iminoglycine Transport Systems; a Study of Transport Specificity

Kidney

Our group has examined the manner in which the two natural iminoacids L-proline and 4-hydroxy-L-proline, and the optically inactive α-amino acid, glycine, are transported across the cellular membranes of mammalian kidney. We have used chemical and kinetic probes (Wilson and Scriver, 1967; Scriver and Wilson, 1967; Scriver, 1968; Mohyuddin and Scriver, 1968; Mohyuddin and Scriver, 1970) both *in vitro* and *in vivo*, in different species; we have also found a useful genetic probe in the mutant phenotype known as hereditary (renal) iminoglycinuria (Scriver and Wilson, 1967; Scriver, 1968). More recently, we have used the normal postnatal ontogeny of the iminoglycine system to reveal additional information as we shall illustrate shortly.

The characteristics of the iminoglycine transport systems in mammalian kidney are summarised in Table 3.1. The transports which operate in the physiological range of substrate concentration are divided into a 'common' system shared by all three compounds, and two other systems, one with specificity for the iminoacids which excludes glycine, and the other serving glycine transport exclusively. The two latter systems have low-Km values (or high affinity as described by the term 1/Km) and low capacity (described by the terms,

TABLE 3.1

The Characteristics of Membrane Transport Systems for Iminoacids and Glycine in Human and Rat Kidney[a]

	L-Proline Systems		*Glycine Systems*	
	'Low-Km'	*'High-Km'*	*'Low-Km'*	*'High-Km'*
Preferred substrate conc. (mM)				
in Man[b]	<0.2	>0.2	<0.1	>0.1
in the Rat[c]	<0.25	>0.25	<0.02	>0.02
Relative capacities				
in Man	1	15	1	>10
in the Rat	1	18	1	8
Partition of uptake (%) between systems at 0.2 mM substrate:				
in Man	60	40	50	50
in the Rat	50	50	30	70
Substrate Preference (specificity)				
in Man	Shared with hydroxy-proline; excludes glycine	Shared with hydroxy-proline and glycine	Excludes imino-acids	Shared with imino-acids
in the Rat	Same as human	Same as human	Same as human	Same as human

[a] Taken from Mohyuddin and Scriver (1970).
[b] Physiological concentrations in plasma of Man: L-proline 0.1-0.4 mM, glycine 0.1-0.5 mM.
[c] Physiological concentrations in plasma of Rat: L-proline about 0.2-0.3 mM, glycine about 0.3-0.4 mM.

Vmax, *in vitro* and Tm, *in vivo*), while the common system exhibits high-Km values for its substrates and a high transport capacity (Mohyuddin and Scriver, 1970). It is the latter system which is apparently inactive in the inborn error of membrane transport, hereditary renal iminoglycinuria. There is also evidence that additional systems serve iminoglycine transport at very high substrate concentrations (Hillman *et al.*, 1968; Hillman and Rosenberg, 1969).

Influence of Tissue Preparation on Kinetic Analysis

Isolated proximal tubule preparations from rabbit kidney have been used by Rosenberg's group to study iminoglycine transport (Hillman *et al.*, 1968; Hillman and Rosenberg, 1969, 1970). These preparations are more homogeneous than the cortex slices which we use in our work; moreover the tubule preparations were examined under conditions of initial-rate kinetics rather than the steady-state kinetics with which we examined slices. Yet both approaches yield data about the chemical and kinetic specificity of iminoglycine transport in the physiological range which are essentially complementary. The Yale group has recently been able to isolate tubule membranes (Hillman and Rosenberg, 1970) to study the binding specificity for L-proline; again their findings parallel our own data obtained from slice preparations. These findings indicate that the rather simple slice experiments provide useful information which can be confirmed by more complex techniques designed to study membrane phenomena specifically.

Iminoglycine Transport in Tissues Other Than Renal Tubule

Other tissues have been examined for the presence of the various iminoglycine transport systems, in order to inform us about the expression of transport gene products in different tissues. Mackenzie in our laboratory (Mackenzie and Scriver, 1970) has found high-Km and low-Km proline transport systems which operate in the physiological range in isolated rat glomerulus. The glomerulus, which is of more complex embryological origin than the proximal renal tubule, has a very special major transport function, namely filtration; nonetheless, some of its membranes must contain the same transport gene products that are active in tubule membranes.

On the other hand, peripheral leukocytes reveal a different pattern of transport sites. The mutant iminoglycinuria phenotype is not expressed in these cells (Tada *et al.*, 1966), suggesting absence of the high-Km system. However, it is possible that the Japanese workers did not actually test for the presence or absence of this system since they did not use high concentrations of substrate.

Reticulocyte membranes are armed initially with high-Km and low-Km systems but during differentiation to mature RBC, they lose their high-Km systems (Winter and Christensen, 1965).

Intestinal absorptive membrane has been examined extensively for the presence of the high-Km systems and a shared system for N-substituted amino acids was found (Lin *et al.*, 1962; Hagihira *et al.*, 1962); however an effort has not been made to identify the equivalent low-Km systems. Preliminary evidence from our group suggests that there are two systems for L-proline uptake in this tissue also.

Brain is another organ with active transport systems; these are deployed in a manner which mimics the kidney systems and which serve uptake at high physiological concentrations (Blasberg and Lajtha, 1965). Data concerning specificity at low substrate concentrations are still scarce in this tissue.

The Use of Ontogeny as a Probe of Transport Specificity

A number of workers, for example, Deren and colleagues (1965), Christensen's group (Winter and Christensen, 1964), Webber's group (Webber and Cairns, 1968; Webber, 1968) and Segal's group (States and Segal, 1968; Segal and Smith, 1969), have examined the ontogeny of various transport systems in mammalian tissues. It is apparent that the activity of membrane transport proteins can vary during cellular differentiation. Once again, the mammalian kidney is a good source of such information, and we set out to examine the iminoglycine system in the newborn to determine how changes in its transport activity, during postnatal development are accomplished. We asked the question whether augmented tubular transport activity during maturation reflects an increase in the tubular surface area which is available for transport, an enhanced coupling of energy to existing transport sites, or a change in the specific activity of transport sites representing appearance of these proteins in the membrane.

Renal iminoglycinuria occurs in early human infancy (Woolf and

TABLE 3.2

Ontogeny of Iminoglycine Transport in Rat Kidney Cortex Slices: Distribution ratios[a]

	Five-minute[b] *Isotope Distribution Ratio*				*Ninety-minute*[b] *Isotope Distribution Ratio*			
	L-proline		*glycine*		*L-proline*		*glycine*	
Age of animal after birth	0.051 mM	12 mM	0.027 mM	6.0 mM	0.051 mM	12 mM	0.027 mM	6.0 mM
1 week	0.33	0.58	0.57	0.52	4.11	2.2	8.37	3.5
2 weeks	0.78	0.53	0.83	0.60	3.37	2.3	9.10	2.8
3 weeks	0.91	0.75	0.68	0.75	2.68	2.4	7.65	2.8
Adult	1.14	0.70	1.36	0.68	3.02	2.0	7.64	2.7

[a] ^{14}C Isotope distribution ratios indicate soluble cpm/ml intracellular water: cpm/ml extracellular fluid. Counts lost as $^{14}CO_2$ (Mohyuddin and Scriver, 1970) are not included in calculation of the ratios presented in the table.

[b] Five-minute uptake is equivalent to 'initial rate'; ninety-minute uptake is equivalent to the 'steady state' and when the transport attributable to the non-saturable components is unity. Data extended from Baerlocher *et al.* (1970).

Norman, 1957; Brodehl and Gellisen, 1968) but it disappears about six months after birth. We found that renal iminoglycinuria also characterised the early infancy of other mammalian species (Baerlocher *et al.,* 1970); again, the rat offered an excellent opportunity for investigation of this problem. We used the cortex slice method for this work, taking care to determine the effect of age upon cellular fluid spaces and the effect of slice size and geometry on substrate uptake. Uptake of iminoacids and glycine in newborn kidney was characterised by the following (Baerlocher *et al.,* 1970):

1. Initial rates of uptake were lower in newborn kidney compared to the mature kidney; this was true primarily at low substrate concentrations (Table 3.2).

2. Steady-state uptake was higher in the newborn kidney than in the mature kidney. This was true at high and low substrate concentrations (Table 3.2).

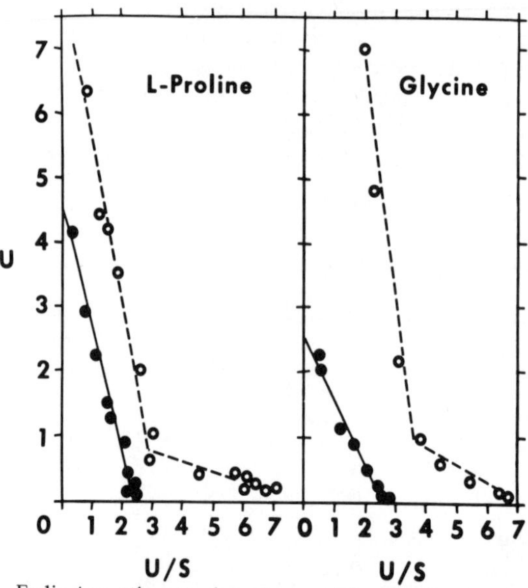

Fig. 1. Eadie-Augustinsson plots show uptake kinetics for L-proline and glycine in one-week-old (solid) circles) and adult (open circles) rat kidney cortex slices. One system with a Km value of about 2 mM serves proline uptake in the newborn; one system also serves glycine uptake (Km about 1 mM). Mature kidney has two systems for uptake of each amino acid; low-Km systems for proline and glycine uptake (Km values of each about 0.1 mM) have appeared. High-Km uptake of proline and glycine in the newborn and adult is achieved on a common system (from Baerlocher *et al.,* 1970).

3. Low initial uptake rates at low substrate concentrations in newborn rat kidney, were explained by the absence of the low-Km systems (Fig. 1). The low-Km proline system emerges at about one week after birth; the low-Km glycine system appears during the third week after birth (Fig. 2). The predicted inhibition profiles, compatible with the absence of low-Km systems which can exclude inhibitors of

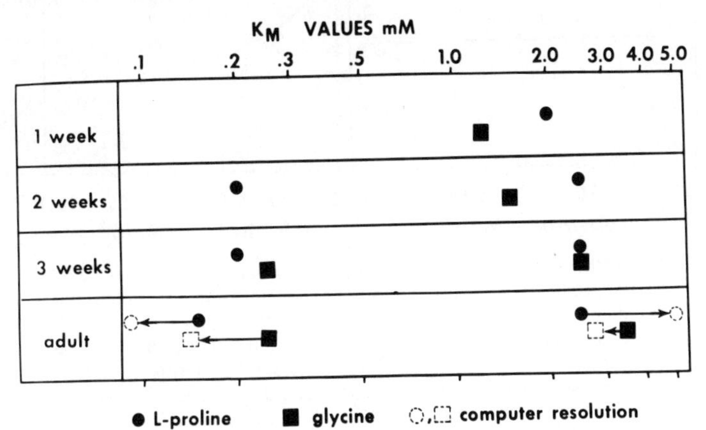

Fig. 2. A graph to illustrate Km values for uptake of proline and glycine at different ages in rat kidney cortex slices from newborn and adult animals. The low-Km proline system appears in the second week; the low-Km glycine system appears in the third week after birth. Interrupted symbols for adult, represent values resolved by iterative analysis, assuming simultaneous uptake on more than one system as shown by Mohyuddin and Scriver (1970).

uptake at low substrate concentrations, were identified (Fig. 3). These findings indicate independent activation of transport sites.

4. The enhanced retention of free amino acids by newborn kidney, reflected in the high steady-state distribution ratios, was explained by a reduced rate of efflux (Table 3.3). Since this affected the internal substrate when preloaded at both high and low concentrations, we must conclude that the specificity of the efflux carrier is not only under different ontogenetic control but that it exhibits different specificity from influx sites.

Transport Ontogeny and Metabolism of Substrate

We also examined the relation of transport ontogeny to intracellular metabolism. We had not anticipated that the low-Km transport systems,

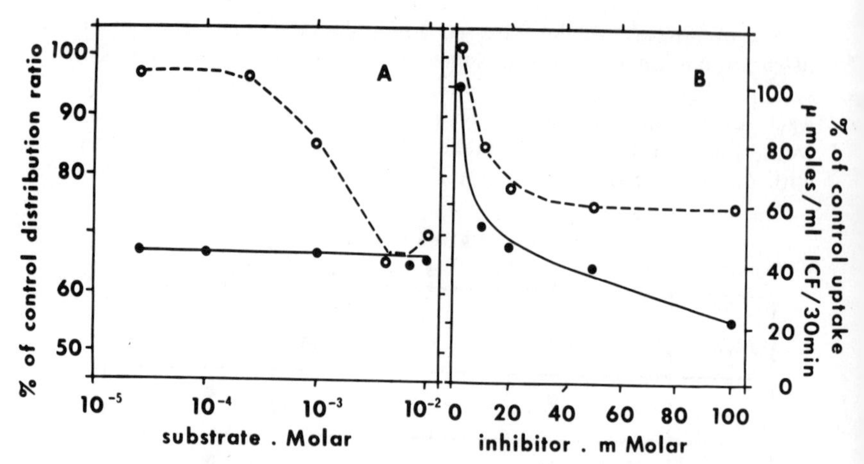

Fig. 3. Inhibition of L-proline uptake by α-aminoisobutyric acid (AIB) under steady-state conditions in newborn (one-week-old) (solid circles) and adult (open circles) rat kidney cortex slices.

A: constant-ratio test; inhibitor : substrate, 10:1. B: variable-ratio test; substrate conc. = 0.2 mM. AIB inhibits L-proline uptake more completely in newborn kidney, indicating absence of a specific (low-Km) system for proline uptake, which excludes AIB. Inhibition expressed as percent of proline uptake by control slices taken from same animal but not exposed to AIB.

TABLE 3.3

Ontogeny of Efflux of L-proline and Glycine from Rat Kidney Cortex Slices

	Percent of radioactivity remaining in slice at 10 min			
Age of animal after birth	L-proline		Glycine	
	0.05 mM	12 mM	0.03 mM	6 mM
1 week	55	50	76	66
2 weeks	48	44	71	50
3 weeks	44	43	–	47
Adult (150-200 g)	34	30	45	35

Efflux measured as described previously (Baerlocher *et al.*, 1970). The slices from different animals were preloaded at a given external concentration until the same amount of radioactivity had been taken up internally. Efflux was then measured when the rate was exponential between 3 and 20 minutes, after transfer of slices into the efflux medium without amino acid.

which are perhaps the more important for cellular economy, would be deficient after birth, when growth must be supported by an efficient amino acid supply. What function does this particularly ontogenetic process then serve?

We observed that incorporation of labelled L-proline and glycine into insoluble material (presumably protein) is greater in newborn kidney than in mature kidney, while oxidation to CO_2 is less in the newborn (Fig. 4). Reduced oxidation rates, combined with the greater intracellular retention of soluble substrate, would assure a substantial intracellular pool to support protein synthesis during growth. Thus, renal tissue has a mechanism which compensates for the absence of the low-Km systems. One must then ask: do these ontogenetic metabolic relationships themselves influence the specific activity of transport sites? The answer is no, because although metabolism of intracellular

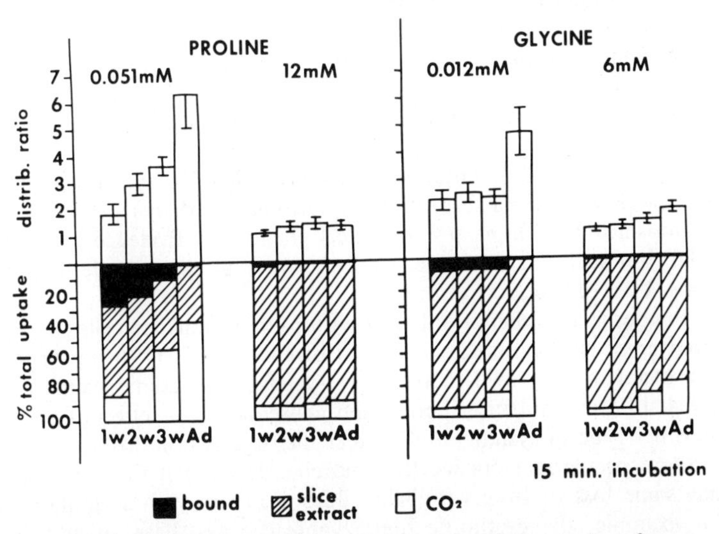

Fig. 4. Distribution of ^{14}C in various intracellular pools, after 15 min incubation kidney cortex slices taken from animals of various ages. Upper bars: total isotope distribution ratio (ICF/ECF). Lower bars: distribution of isotope between insoluble (protein), soluble (free proline and metabolites), and volatile (CO_2) pools. Incorporation is greater but catabolism and efflux is less in newborn kidney; consequently the distribution ratio of soluble (free) label is higher (see text and Table 3.1). However, total uptake/ml ICF is less at low external substrate concentrations in the immature kidney because of the deficient influx systems.

substrate modifies the estimation of absolute uptake rate (Mohyuddin and Scriver, 1970), measurements of the affinity of the transport site, and in turn the recognition of absence or presence of a carrier during ontogeny, are not modified by intracellular metabolic events acting on the substrate. This was confirmed by additional studies with the metabolically inert substrate α-aminoisobutyric acid (Baerlocher *et al.*, 1970).

Phenotypic Effects of Transport Mutations

The differential ontogeny of the iminoglycine transport systems apparently does not impair one important function of this membrane system, namely to deliver and maintain an adequate intracellular pool of free amino acids. Is this also true of the inborn errors of amino acid transport? Some of these mutations such as Hartnup disease, cystinuria, and hereditary iminoglycinuria apparently affect only the relevant high-Km system (Scriver and Hechtman, 1970). If the corresponding low-Km systems are still operative then amino acids can still be taken up efficiently by the cell under normal physiological conditions. Few phenotypic manifestations and relatively little effect upon fitness would be anticipated in this circumstance, and in general this is the case.

On the other hand, mutations which affect low-Km systems (such as the methionine malabsorption syndrome perhaps) should be accompanied by a more severe clinical phenotype. Evidence is limited so far, but such phenotypes generally seem to be more severe than the former class of transport disorders.

Obviously such phenotypic relationships in mammalian cells are only a matter of speculation at present. However, Képès (1964) proposed that high-Km and low-Km transport systems in bacterial membranes served different nutritional roles and consequently deletion of one or the other type of system would affect the organism in different ways.

Such speculations are worth some consideration if they remind us that some loss of fitness probably does accompany these mutations. For example, the cystinuric homozygote can also have an intestinal transport defect (Rosenberg *et al.*, 1966). Such an individual will be at great risk if another illness occurs to compromise the efficiency of the remaining intestinal transport systems.* The poor clinical prognosis of cystinuric children who develop the coeliac syndrome (Fleming *et al.*, 1963) is of interest in this context. The absorptive area of the intestine

* These systems may be the low-Km systems as suggested here or dipeptide systems as suggested by Milne (see discussion).

is reduced by the coeliac condition and the efficiency with which the low-Km systems operate in the cystinuric subject will be impaired in the cystinuric patient. This would greatly impair nutrition of the essential amino acid lysine and may explain the severe growth failure of such patients.

One can predict that iminoglycinuric homozygotes will have little or no tubular reabsorption of iminoacids and glycine in the immediate postnatal period, because the high-Km system is deleted by the mutation and the low-Km systems are not yet operative. Levy (personal communication, 1970) has now observed several hereditary iminoglycinurics in the newborn period. The iminoglycinuria is extraordinarily heavy in these infants, apparently representing almost the complete filtered load of the iminoacids and glycine.

A Cerebral Phenotype in Cystinuria?

Such observations suggest that certain types of gene-environment interactions during growth and development may place the individual with a transport mutation at grave risk and that there may be an undesirable outcome of such interaction. We have recently identified a phenomenon which illustrates this concern (Scriver *et al.*, 1970). Cystinuria is generally considered to be a trait with renal and intestinal phenotypes (Rosenberg *et al.*, 1966) but with little or no other clinical manifestations. However, in our Montreal survey of hereditary aminoacidopathies among 1400 patients with psychiatric illness or mental retardation, we found two young schizophrenic patients with homozygous cystinuria; there were no other inborn errors identified in this population. We found the frequency of homozygous cystinuria in 4700 members of the general population to be much lower, on the order of 10^{-4}. Cystinuria was thus about 10 times more frequent in patients with mental illness than in the general population. Our observations are supported by those of other workers. Berry (1962), Carson (personal communication, 1970), Efron (1965) and Visakorpi and Hyrske (1960), have all observed a high frequency of cystinuria in hospitals for the mentally ill or retarded. On the other hand, Levy and colleagues (1968, and personal communication, 1970) have found the frequency of cystinuria at birth to be about 10^{-4} in Massachusetts infants screened for hyperaminoacidurias. Statistical analysis of the frequency rates in all these studies (Fig. 5) reveals the difference between psychiatric and non-psychiatric population to be significant.

The foregoing implies that patients with cystinuria are at greater risk than non-cystinurics to develop mental illness; in other words, there appears to be a 'cerebral phenotype' in some patients with

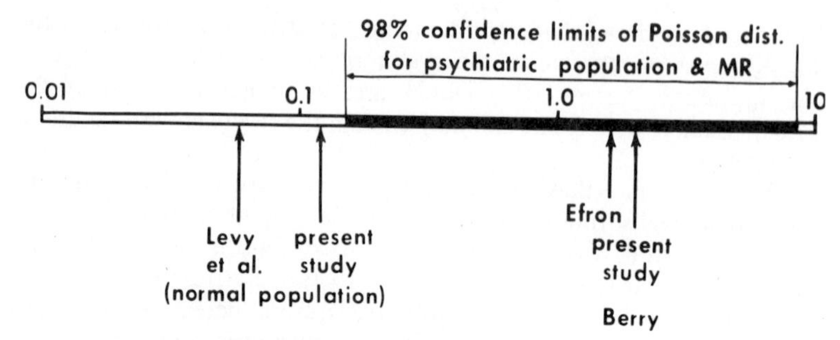

Fig. 5. Frequency of cystinuria in given population of 1400 persons. The graph shows on a logarithmic scale the number of patients found in psychiatric hospitals according to Scriver *et al.*, (1970) (present study) and Efron (1965). The black portion of the graph represents 98% confidence limits of a Poisson distribution based on these observed frequencies. Berry (1962), Carson (1970), and Visakorpi and Hyrske (1960) all observed the same, or higher frequency, of cystinuria in patients with mental illness.

The frequency of cystinuria in a general hospital population of 1400 subjects is considerably lower (present study). The same is true for the observed frequency of cystinuria at birth (Levy *et al.*, 1968; personal communication, 1970). Both of the latter frequencies fall outside the 98% confidence limits of the Poisson distribution. The frequency of cystinuria is thus truly about 10 times higher in patients with mental illness, implying that this inborn error of membrane transport places the individual at a greater risk than usual to develop mental illness.

homozygous cystinuria. If there are environmental conditions which bring this out, it will be important to discover what they are, so that they can be avoided. If on the other hand, the cerebral phenotype reflects yet another genotype for cystinuria, we must know this too; it can influence our counselling of persons carrying the allele.

Summary

Chemical, kinetic and genetic probes of plasma membrane permeability reveal that a mosaic of transport sites (proteins) serves the flux of organic solutes into and out of the cell. A study of the ontogeny of the 'iminoglycine' transport systems in mammalian kidney, shows that influx and efflux permeabilities evolve independently, and that each of the three physiological systems for iminoacid and glycine influx is apparently under independent genetic control.

Mutations which affect transport sites have phenotypic effects which will depend on the system affected, and the overall significance of this system for cellular nutrition. The inborn errors of amino acids transport of man can and apparently do affect development and function of the cerebral nervous system under particular conditions of gene-environment interaction.

REFERENCES

Baerlocher, K., Scriver, C. R. and Mohyuddin, F. (1970). *Proc. Nat. Acad. Sci.* **65,** 1009.

Berlin, R. D. (1970). *Science* **168,** 1539.

Berry, H. K. (1962). *Amer. J. Mental Defic.* **66,** 555.

Blasberg, R. and Lajtha, A. (1965). *Arch. Biochem. Biophys.* **112,** 361.

Brodehl, J. and Gellisen, K. (1968). *Ped.* **42,** 395.

Carson, N. A. J. (1970). Personal communication.

Christensen, H. N. (1967). *Perspect. Biol. and Med.* **10,** 471.

Christensen, H. N. (1969). *Adv. in Enzymol.* **32,** 1.

Collander R. and Barlund, H. (1933). *Acta Botanica Fennica* **11,** 1.

Danielli, J. F. and Davson, H. (1935). *J. Cell Comp. Physiol.* **5,** 495.

Deren, J. J., Strauss, E. W. and Wilson, T. H. (1965). *Develop. Biol.* **12,** 467.

Efron, M. L. (1965). *New Eng. J. Med.* **272,** 1058 and 1107.

Fleming, W. H., Avery, G. B., Morgan, R. I, and Cone, T. E. (1963). *Pediatrics* **32,** 358.

Hagihira, H., Wilson, T. H. and Lin, E. C. C. (1962). *Amer. J. Physiol.* **203,** 637.

Hillman, R. E., Albrecht, I. and Rosenberg, L. E. (1968). *J. Biol. Chem.* **243,** 5566.

Hillman, R. E. and Rosenberg, L. E. (1969). *J. Biol. Chem.* **244,** 4494.

Hillman, R. E. and Rosenberg, L. E. (1970). *Biochim. Biophys. Acta.* **211,** 318.

Képès, A. (1964). In 'The Cellular Function of Membrane Transport' (ed. J. F. Hoffman), p. 155. Englewood Cliffs, N. J.: Prentice-Hall.

Korn, E. D. (1969). *Fed. Proc.* **28,** 6.

Levy, H. L., Shih, V. E., Madigan, P. M., Karolkewicz, V. and MacCready, R. A. (1968). *Clin. Biochem.* **1,** 208.

Lin, E. C. C., Hagihira, H. and Wilson, T. H. (1962). *Amer. J. Physiol.* **202,** 919.

Mackenzie, S. and Scriver, C. R. (1970). *Biochim. Biophys. Acta* **196,** 110.

Mohyuddin, F. and Scriver, C. R. (1968). *Biochem. Biophys. Res. Comm.* **32,** 852.

Mohyuddin, F. and Scriver, C. R. (1970). *Amer. J. Physiol.* **219,** 1.

Pardee, A. B. (1968). *Science* **162,** 632.

Robertson, J. D. (1964). In 'Cellular Membranes in Development' (ed. M. Locke), pp. 1-82. New York: Academic Press.

Rosenberg, L. E. (1969). In 'Biological Membranes' (ed. R. M. Dowben), pp. 255-295. Boston: Little, Brown and Co.

Rosenberg, L. E., Downing, S., Durant, J. L. and Segal, S. (1966). *J. Clin. Invest.* **45,** 365.

Schultz, S. C. (1969). In 'Biological Membranes' (ed. R. M. Dowben), pp. 59-108. Boston: Little, Brown and Co.

Scott, D. H., (1891). *Nature* **44,** 580.

Scriver, C. R. (1968). *J. Clin. Invest.* **47,** 823 (No. 4 April).

Scriver, C. R. (1969). *Pediatrics* **44,** 348.

Scriver, C. R. and Hechtman, P. (1970). *Adv. in Hum. Genet.* **1,** 211. New York: Plenum Press.

Scriver, C. R., Whelan, D. T., Clow, C. L. and Dallaire, L. (1970). *New Eng. J. Med.* **283,** 783.

Scriver, C. R. and Wilson, O. H. (1967). *Science* **155,** 1428.

Segal, S. and Smith, I. (1969). *Biochem. Biophys. Res. Comm.* **35,** 771.

States, B. and Segal, S. (1968). *Biochim. Biophys. Acta* **163,** 154.

Tada, K., Morikawa, T. and Arakawa, T. (1966). *Tohoku, J. Exp. Med.* **90,** 189.

Visakorpi, J. K. and Hyrske, I. (1960). *Ann. Ped. Fenn.* **6,** 112.

Webber, W. A. (1968). *Canad. J. Physiol. Pharmacol.* **46,** 765, (No. 5. Sept.).

Webber, W. A. and Cairns, J. A.(1968). *Canad. J. Physiol. Pharmacol.* **46,** 165.

Wilson, O. H. and Scriver, C. R. (1967). *Amer. J. Physiol.* **213,** 185.

Winter, C. G. and Christensen, H. N. (1965). *J. Biol. Chem.* **240,** 3594.

Woolf, L. I. and Norman, A. P. (1957). *J. Pediat.* **50,** 271.

Defects in Renal Tubular Reabsorption of Amino Acids

D. N. RAINE

Biochemistry Department, The Children's Hospital, Birmingham, B16 8ET

The evolution of knowledge of the ways in which substances are transported across membranes has a fascinating history and provides many lessons in epistemology. Concepts, such as that of carrier substances, which were once denied or hotly debated, have given way to evidence and the actual carrier molecules are now being isolated (Langridge *et al.*, 1970). The belief that a given substance might be transported both by a specific mechanism and by one which handled it and other related substances as a group went someway to explain, otherwise inexplicable, experimental observations. The determinants of the direction of transport into or out of the cell or into or out of the animal system across the two membranes of a layer of cells are still being explored.

There is no doubt, however, that transport processes do exist and that they are capable of functioning less efficiently than usual or even failing completely. In a number of instances such malfunction does not appear to have any serious consequences but in others the abnormality is associated with quite definite clinical disease.

Some Renal Tubular Transport Systems

For the present it is convenient to consider five group transport systems for amino acids even though one of these, that for the basic amino acids and cystine, is known to be more complex. The five systems involve the following groups of amino acids.

1. *Neutral amino acids*	Leu	$AspNH_2$	Ser	Phe	Gly	CysH
	Ileu	$GluNH_2$	Thr	Tyr	Ala	Meth
	Val			Try		His
						Citrulline
2. *Basic amino acids and cystine*	Cys		Lys	Arg	Orn	
3. *Acidic amino acids*	Asp and Glu					

4. *Imino acids and glycine* Pro Hypro and Gly
5. *β-amino acids* β-Ala, β-aminoisobutyric acid and taurine

Interference with Tubular Transport

Abnormalities of all but the third of these systems have been demonstrated in human subjects and the first two and the last are associated with well defined clinical disorders. Some or all of the transport mechanisms are interfered with in other circumstances however. For example some drugs and chemicals intoxicate the renal tubular epithelium and the products that accumulate in some inherited metabolic diseases seem to exert a similar effect. Furthermore, the absence of some vitamins, which presumably act as essential cofactors, results in a less than normal efficiency of the renal tubular epithelium.

In most cases these effects lead to a fairly generalised amino aciduria but, where there has been the opportunity to follow the process, a progressively more complex amino aciduria is seen to develop and, where the offending agent can be withdrawn, the abnormality is corrected in a similar but reversed sequence.

An understanding of the transport processes also helps to explain certain less generalised forms of amino aciduria associated with specific inherited metabolic disorders which lead to the accumulation in the blood of a single amino acid. Such a situation presents a special challenge to the tubular group transport systems as it tends to overload these, with the result that the other amino acids carried by the same mechanism are dealt with less completely and although their concentration is not abnormally high they are excreted in greater than normal amounts.

Finally, this last situation has been turned to possible advantage in treating metabolic disorders in which a single amino acid accumulates in the blood. The excretion of the amino acid can be facilitated by blocking its reabsorption by administering an innocuous amino acid known to be handled by the same reabsorption mechanism.

In the discussion that follows many disorders of tubular transport will be referred to but descriptions of those not associated with affections of the nervous system will be curtailed.

Conditions Associated with Non-specific Tubular Damage

Any condition damaging the renal tubular epithelium will affect its transport functions, including those for amino acids, and lead to a greater or lesser degree of amino aciduria. Several diseases associated with such a renal amino aciduria, for example deficiency of vitamin D, leading to *rickets* (Jonxis and Huisman, 1953), and the *'Fanconi syndrome'* do not notably affect the nervous system or lead to intellectual impairment. The amino aciduria only occurs when vitamin D deficiency has stimulated secondary hyperparathyroidism: the appearance of rickets usually coincides with this (Fraser *et al.*, 1967). The amino aciduria hitherto believed to be associated with *scurvy* (Jonxis and Huisman, 1954) has recently been questioned (Brodehl and Kaas, 1970).

Other conditions such as *Wilson's disease* (hepatolenticular degeneration) (Bickel *et al.*, 1957) and *galactosaemia* (Holzel *et al.*, 1952), do affect the nervous system and galactosaemia, at least, illustrates how endogenously produced metabolites such as galactose 1-phosphate, whose further metabolism is impaired, can damage the tissue of the lens to produce a cataract (Gitzelmann *et al.*, 1967) and at the same time damage the renal tubular epithelium sufficiently to cause amino aciduria. When galactose is withheld from the patient and the concentration of galactose 1-phosphate falls, the amino aciduria gradually clears but it may take several days to do so (Komrower *et al.*, 1956).

A similar phenomenon is seen when the tubules are poisoned with other substances such as *maleic acid* (Rosenberg and Segal, 1964), *degradation products of tetracycline* (Cleveland *et al.*, 1965) or even Worcestershire sauce (Murphy, 1967). In the case of old tetracycline the first amino acid to appear and the last to become normal after withdrawal of the offending drug is glycine but the full effect is to produce a nearly generalised amino aciduria.

Heavy metals such as *cadmium, copper, uranium* and *lead* are associated with amino aciduria. Usually this is generalised but some amino acids may be affected more than others and in some instances there are associated tubular defects involving phosphorus and protein. Thus, in cadmium poisoning threonine, cystine, citrulline and tyrosine are reported to be prominent (Adams *et al.*, 1969). The amino aciduria in lead poisoning is said to affect β-amino isobutyric acid particularly (Marsden and Wilson, 1955).

Lowe's syndrome perhaps should not be included under the heading of non-specific tubular damage but the real cause of this condition is still to be defined. Meanwhile it does not readily fall into any category. This sex linked disorder is associated with amino aciduria but clinically the electrolyte disturbances that are so easily precipitated by quite minor intercurrent infections are of much greater importance. Mental and physical development are retarded, cataracts and glaucoma affect the vision and apart from the more or less generalised amino aciduria there is also proteinuria (Abbassi *et al.*, 1968), the situation may be further complicated by clinically similar variants, one associated with the excretion of ornithine (Schwartz *et al.*, 1964), one associated with the excretion of cystine and the dibasic amino acids (Jagenburg, 1959) and possibly another variety with more prominent neurological involvement resembling a leucodystrophy (Martin *et al.*, 1967).

Specific Disorders of Tubular Reabsorption

Hartnup Disease

As more cases of this condition are described the more variable does the clinical picture become. Cerebellar ataxia in older children and adults is prominent but varies in its severity. Mental symptoms and hallucinations may or may not be present and also vary in intensity. Intelligence may be impaired but is not so in all cases. A red scaly pellagra-like skin rash on the exposed parts is made worse by exposure to sunlight. Children are not commonly affected in the first years of life and infants are believed to be normal. However, the one case we have studied presented with recurrent diarrhoea and was considered to be intolerant of milk, possibly due to a primary or secondary deficiency of intestinal lactase for some months before the correct diagnosis was reached. One other case in the literature had a similar history but attention was not specifically drawn by the authors to this feature (Halvorsen and Halvorsen, 1963).

The amino aciduria associated with Hartnup disease is remarkable in that it involves more than a dozen species and yet there are several notable exceptions if it is compared with the generalised amino aciduria of liver disease or of heavy metal poisoning. The exceptions are neatly explained in terms of the group transport mechanisms: the acidic amino acids, the dibasic amino acids, the imino acids and glycine are all unaffected. The amino acids excreted are all in the neutral group although two members of this group, cystine and methionine, are not excreted and it is not clear at present why this should be.

Hartnup disease illustrates a nearly general rule that, where renal tubular transport is abnormal, intestinal transport also can be shown to be defective. Occasionally the defects in transport in these two tissues differ and this is due to the fact that the intestinal and tubular epithelium do not possess exactly the same complement of transport mechanisms.

The precise explanation of the symptomatology of Hartnup disease in terms of the biochemical defect is still speculative. The pellagra-like skin rash is believed to be due to nicotinamide deficiency resulting from the limited formation of this vitamin from tryptophan which fails to be absorbed in sufficient quantity from the intestine. This same deficiency of nicotinamide may explain the cerebellar symptoms, for these are also seen with the skin rash in a different condition, congenital tryptophanaemia (Tada *et al.*, 1963) in which there is also a failure to convert tryptophan to nicotinamide for a metabolic reason rather than one of malabsorption. A secondary consequence of the failure in absorption is that bacterial fermentation takes place, the products of which may cause the diarrhoea seen in our case and perhaps some other symptoms, but which probably accounts for most of the secondary biochemical abnormalities seen in the urine and, to some extent, also in the faeces. Many of these problems are discussed more fully by Professor Jepson in the present symposium.

Cystinuria

This condition, usually only associated with the formation of urinary calculi and then only in a proportion of affected subjects, illustrates how a series of observations, at first explained by a simple and well accepted theory, can call ultimately for a more complex explanation differing substantially from that first proposed. The finding that cystine was accompanied by only the three dibasic acids in urine led Dent and Rose (1951) to suggest that the disease was due to defective reabsorption of these four amino acids by a specific group transfer mechanism in the renal tubule and that, because of its relative insolubility, cystine was deposited in the renal tract as calculi. This form of the disease, which is not associated with mental defect, is now known to exist in three phenotypes, one associated with a similar defect in the transport of these same amino acids across the intestine (Rosenberg *et al.*, 1966).

As already indicated, the management of cystine and the dibasic acids is probably more comlex than has hitherto been suspected. A condition associated with the excretion of only the three dibasic acids has been described by Whelan and Scriver (1968). There is some evidence that classical cystinuria may be a disorder of efflux of the four amino acids rather than one of tubular reabsorption. These concepts are still being developed and are discussed more fully by Dr. Scriver in the present volume and in another shortly to be published (Scriver and Whelan, 1971).

Imino-glycinuria

The history of this abnormality reminds us that not all associated features of disease are causally related. Patients excreting proline, hydroxyproline and glycine were found to have a variety of unrelated symptoms and one collection of these has several times been referred to as Joseph's syndrome (Joseph *et al.*, 1958). However, the number of clinically affected subjects was soon far outweighed by the number of perfectly healthy sibships in whom the same chemical abnormality could be demonstrated (Scriver, 1968). This transport mechanism will be referred to again but defects in it are not believed to have any clinico-pathological significance.

Group Transport Saturation in Single Amino Acidopathies

An important part of the study of group transport processes has been the effect of infusion of an excessive amount of one of the members of the group on the excretion of the others. Usually if the transport mechanism is overloaded by one of its passengers the others fail to be reabsorbed as efficiently as usual and they therefore appear in the urine. Indeed careful interpretation of the quantitative aspects of these experiments allows the affinity of the carrier mechanism for each of the molecular species with which it is concerned to be calculated.

These same experiments sometimes occur naturally in the form of inherited metabolic diseases in which a single amino acid accumulates in the blood and is filtered by the glomerulus, thus presenting an unusual load to the renal tubule. Several examples are known.

Hyperprolinaemia

There are two forms of this disease due to different enzyme deficiencies. In one patient the brain and some facial bones were malformed. More usually, however, the history is of progressive mental retardation and fits, although intelligence is not always impaired (Schafer *et al.*, 1962). This is specially true of the second type of the disease (Emery *et al.*, 1968). Many patients have been deaf.

Because the concentration of proline in the blood may be several times the normal value this amino acid is excreted to such an extent as to overload its tubular reabsorptive mechanism. Since proline is transported in association with hydroxyproline and glycine, it is not surprising that these two amino acids are also found in abnormal quantities in the urine of patients with prolinaemia.

β-Alaninaemia

This condition has only rarely been described (Scriver *et al.*, 1966) but it illustrates the same principle as does prolinaemia but for the mechanism transporting β-amino acids. Here, as a result of an enzyme deficiency, excessive amounts of β-alanine are filtered and the fact that its excretion is accompanied by greater than usual amounts of β-amino isobutyric acid, taurine and γ-aminobutyric acid demonstrates that these amino acids share a common group transport process.

The condition itself is a serious one with somnolence and convulsions followed by death in infancy. The pathogenesis is not yet understood.

Citrullinuria

This condition, due to a defect in argininosuccinate synthetase, one of the several enzymes in the urea cycle, again has marked central nervous system involvement. The onset of clinical symptoms has been delayed for a few months or even years but attacks of vomiting with an associated alkalosis proceed to a fluctuating series of fits, somnolence and coma. Other neurological signs such as hypotonia, tremor and exaggerated reflexes have been observed (McMurray *et al.*, 1963).

Citrulline is one of the neutral amino acids and when the group transport mechanism for these is presented with an unusual amount of citrulline the reabsorption of alanine, glycine, histidine and serine is reduced and these amino acids appear in greater than normal amounts in the urine. The fact that other neutral amino acids such as the branched-chain acids and the aromatic acids are not affected may reflect the relative affinities of these for the carrier mechanism compared with those whose reabsorption is impaired. A case of citrullinuria associated with the excretion of cystine and the three basic amino acids described by Visakorpi (1962) is not easily understandable in terms of the tubular transport mechanisms and this may be a coincidental occurrence of two defects in the same subject.

Argininaemia

Another example of a single amino acid overloading a tubular transport mechanism as a result of an inherited metabolic disease, also involves a defect in a urea cycle enzyme, this time arginase (Terheggen *et al.*, 1969). As would be expected the high concentration of arginine filtered from the blood results in the excretion of this amino acid accompanied by cystine, lysine and ornithine. The clinical symptoms in this case include severe mental retardation, spastic diplegia and convulsions and are sufficient to differentiate the disease from cystinuria in which the same urinary abnormalities occur.

Saturation of Transport Systems in Therapy

Following the discovery that phenylketonuria could be treated by a diet low in phenylalanine, the specific amino acid affected by the genetic enzyme defect in this disease (Bickel *et al.*, 1953), a dietary approach has been used in the treatment of similar disorders such as branched-chain amino aciduria (maple syrup urine disease) and histidinaemia. Metabolic disorders involving non-essential amino acids have not proved so amenable to this approach as their main source has been endogenous rather than dietary.

An alternative method of reducing an abnormally high plasma concentration is to facilitate the excretion of the offending substance by inhibiting its reabsorption in the renal tubule after it has been filtered by the glomerulus. The feasibility of this has been explored in the case of homocystinuria (Cusworth and Gattereau, 1968, Cusworth and Dent, 1969) and in hydroxyprolinaemia (Raine, 1969). In the case of homocystinuria (homocystine is carried by the cystine-dibasic amino acid transport mechanism) infusions of lysine and of arginine glutamate promoted the excretion of homocystine. Oral administration, however, appeared to have too little effect to be of long term therapeutic value.

Our patient with hydroxyprolinuria however, who had a plasma hydroxyproline concentration of 0.5 mmol/l has had this halved by adding 10 g glycine to each of her three main meals in the day. Rather than increase the glycine load further on a long term basis the possibility that the plasma hydroxyproline concentration may be still further reduced by blocking the transport mechanism by adding proline to the glycine load is being explored. Such an approach may be more general in its application than this and the possible value of synthetic substances, capable of more specific and effective inhibition of these transport mechanisms, should be considered for the future.

REFERENCES

Abbassi, V., Lowe, C. U. and Calcagno, P. L. (1968). Oculo-cerebro-renal syndrome. *Amer. J. Dis. Child.* **115,** 145.

Adams, R. G., Harrison, J. F. and Scott, P. (1969). The development of cadmium induced proteinuria, impaired renal function and osteomalacia in alkaline battery workers. *Quart. J. Med.* **38,** 425.

Bickel, H., Gerrard, J. and Hickmans, E. M. (1953). Influence of phenylalanine intake on phenylketonuria. *Lancet* **2,** 812.

Bickel, H., Neale, F. C. and Hall, G. (1957). A clinical and biochemical study of hepatolenticular degeneration (Wilson's disease). *Quart J. Med.* **26,** 527.

Brodehl, J. and Kaas, W. P. (1970). Tubular reabsorption of free amino acids in infants with scurvy. *Clin. chim. Acta* **28,** 409.

Cleveland, W. W., Adams, W. C., Mann, J. B. and Nyhan, W. L. (1965). Acquired Fanconi syndrome following degraded tetracycline. *J. Pediat.* **66,** 333.

Cusworth, D. C. and Dent, C. E. (1969). Homocystinuria. *Brit. Med. Bull.* **25,** New Aspects of Human Genetics, 42.

Cusworth, D. C. and Gattereau, A. (1968). Inhibition of renal tubular reabsorption of homocystine by lysine and arginine. *Lancet* **2,** 916.

Dent, C. E. and Rose, G. A. (1951). *Quart. J. Med. N.S.* **20,** 205.

Emery, F. A., Goldie, L. and Stern, J. (1968). Hyperprolinaemia type 2 *J. Ment. Defic. Res.* **12,** 187.

Fraser, D., Kooh, S. W. and Scriver, C. R. (1967). Hyperparathyroidism as the cause of hyperaminoaciduria and phosphaturia in human vitamin D deficiency. *Pediat. Res.* **1,** 425.

Gitzelmann, R., Curtius, H. and Schneller, I. (1967). Galactitol and galactose-1-phosphate in the lens of a galactosemic infant. *Exp. Eye Res.* **6,** 1.

Halvorsen, K. and Halvorsen, S. (1963). Hartnup disease. *Pediatrics* **31,** 29.

Holzel, A., Komrower, G. M. and Wilson, V. K. (1952). Amino aciduria in galactosaemia. *Brit. med. J.* **1,** 194.

Jagenburg, O. R. (1959). The urinary excretion of free amino acids and other amino compounds in the human. *Scand. J. Clin. lab. Invest.* **11,** suppl. 43.

Jonxis, J. H. P. and Huisman, T. H. J. (1953). Amino aciduria in rachitic children. *Lancet* **2,** 428.

Jonxis, J. H. P. and Huisman, T. H. J. (1954). Amino aciduria and ascorbic acid deficiency. *Pediatrics* **14,** 238.

Joseph, R., Ribierre, M., Job, J. C. and Girault, M. (1958). Maladie familiale associant des convulsions à début très précoce une hyperalbuminorachie et une hyperaminoacidurie. *Arch Franç. Pédiat.* **15,** 374.

Komrower, G. M., Schwartz, V., Holzel, A. and Golberg, L. (1956). A clinical and biochemical study of galactosaemia. *Arch. Dis. Child* **31,** 254.

Langridge, R., Shinagawa, H. and Pardee, A. B. (1970). Sulphate binding protein from *Salmonella typhimurium:* physical properties. *Science* **169,** 59.

McMurray, W. C., Rathbun, J. C., Mohyuddin, F. and Koegler, S. J. (1963). Citrullinuria. *Pediatrics* **32,** 347.

Marsden, H. B. and Wilson, V. K. (1955). Lead poisoning in children: correlation of clinical and pathological findings. *Brit. med. J.* **1,** 324.

Martin, L., Martin, J. J., Guazzi, G. C., Lowenthal, A. and Maniewski, J. (1967). New oculo-oto-cerebro-renal syndrome. *Lancet* **1,** 1112.

Murphy, K. J. (1967). Bilateral renal calculi and amino aciduria after excessive intake of Worcestershire sauce. *Lancet* **2,** 401.

Raine, D. N. (1969). Screening for inherited metabolic disease. *Ann. clin. Biochem.* **6,** 29.

Rosenberg, L. E., Downing, S., Durant, J. L. and Segal, S. (1966). Cystinuria: biochemical evidence for three genetically distinct diseases. *J. clin. Invest.* **45,** 365.

Rosenberg, L. E. and Segal, S. (1964). Maleic acid-induced inhibition of amino acid transport in rat kidney. *Biochem. J.* **92,** 345.

Schafer, I. A., Scriver, C. R. and Efron, M. L. (1962). Familial hyperprolinemia, cerebral dysfunction and renal anomalies occurring in a family with hereditary nephropathy and deafness. *New Eng. J. Med.* **267,** 51.

Schwartz, R., Hall, P. W. and Gabuzda, G. J. (1964). Metabolism of ornithine and other amino acids in the cerebro-oculo-rrenal syndrome. *Amer. J. Dis. Child.* **36,** 778.

Scriver, C. R. (1968). Renal tubular transport of proline, hydroxyproline and glycine. III Genetic basis for more than one mode of transport in human kidney. *J. clin. Invest.* **47**, 823.

Scriver, C. R., Pueschel, S. and Davies, E. (1966). Hyper-β-alaninemia associated with β-amino aciduria and γ-aminobutyric aciduria somnolence and seizures. *New Eng. J. Med.* **274**, 635.

Scriver, C. R. and Whelan, D. T. (1971). Cystinuria: concepts and new observations. In 'Some Inherited Disorders of Sulphur Metabolism', (eds N. A. J. Carson and D. N. Raine). Edinburgh: E. and S. Livingstone.

Tada, K., Ito, H., Wada, Y. and Arakawa, T. (1963). Congenital tryptophanuria with dwarfism. *Tohoku J. exp. Med.* **80**, 118.

Terheggen, H. G., Schwenk, A., Lowenthal, A., van Sande, M. and Colombo, J. P. (1969). Argininaemia with arginase deficiency. *Lancet* **2**, 748.

Visakorpi, J. K. (1962). Citrullinuria. *Lancet* **1**, 1357.

Whelan, D. T. and Scriver, C. R. (1968). Hyperdibasic amino aciduria: an inherited disorder of amino acid transport. *Pediat. Res.* **2**, 525.

DISCUSSION OF PAPERS BY DR. C. R. SCRIVER AND ASSOCIATES AND OF A PAPER BY DR. D. N. RAINE

Milne: Have you taken into consideration the possibility that the incidence of consanguinity is likely to be higher than average in patients either with mental retardation or cystinuria so that the two states might co-exist by chance rather than because of cause and effect?

Scriver: Yes. There was no known consanguinity in the parents of the two proven cystinurics amongst 1400 patients of a psychiatric hospital. Both in this survey and in the study by Efron, consanguinity could apparently be excluded.

Allison: Nevertheless even if it is not possible to elicit a history of consanguinity, the possibility remains that there exists a higher likelihood of consanguinity in patients of a mental institution than in the general population.

Komrower: We are carrying out a biochemical screening programme of mentally retarded children and our experience supports the existence of an association between mental retardation and cystinuria. We have found an abnormally high concentration of cystine in the amniotic fluid collected at the delivery of an infant who had demonstrable cystinuria in the first sample of urine which he passed after birth (compared with published normal values). This raises the question as to whether cystinuric foetuses may suffer from cystine deficiency which would predispose to mental subnormality and possibly contribute to the low birth weight which is reported in cystinuric patients.

Scriver: I should tend to incriminate deficiency of lysine even more than cystine since lysine is an essential amino acid.

Benson: It was stated earlier that accumulation of galactose 1-phosphate could give rise to cataracts. However, cataracts develop in subjects with galactokinase deficiency in whom galactose 1-phosphate production is deficient. Since there is an accumulation of free galactose and its derivative galactitol, these may be responsible for cataract formation rather than galactose 1-phosphate.

Jepson: Yes. The phosphate group of galactose 1-phosphate would prevent formation of the alcohol, galactitol, which is probably responsible for the cataracts.

Scriver: The specificity of membrane transport (in humans and in microorganisms) seems to reside in the binding sites on membranes rather than in the process of energy coupling which by comparison may be relatively non-specific. Nonetheless Hechtman and his group discovered a mutant microorganism lacking β-alanine transaminase, which was also unable to retain a pool of β-alanine. He showed that although influx of β-alanine was normal, there was inability to retain an internal concentration of β-alanine. This suggests that energy-coupling to the system was relatively specific in this case.

Yielding: Are the enzyme systems related to membrane transport inducible and are there any high Km enzymes known to develop after birth in humans? Is it known whether postnatal enzyme development is related to gestational age or the process of birth?

Benson: Yes. In a large range of mammalian species there are four types of liver hexokinase demonstrable by starch gel electrophoresis. Three of these exist before birth but type IV or glucokinase, which has a higher Km than the other types cannot be detected until after birth, (e.g. in rats not until about the fourth postnatal week). The stimulus for enzyme development differs with individual enzymes, e.g. premature delivery of rat foetuses results in premature appearance of hepatic glucose 6-phosphatase, phosphopyruvate carboxylase and tyrosine transaminase. On the other hand the development of the arginine synthetase system (which normally occurs shortly after birth) or of the glycogen synthetic enzymes (which normally occurs a few days before birth), appears to be independent of the process of birth but is related to gestational age.

Dr. Scriver and Dr. Komrower have both cited certain evidence that mental subnormality is probably associated with defects in renal tubular reabsorption, such as cystinuria. This is supported by a study carried out by Drs. P. Swift, J. Studdy and myself. We investigated 30 families in which there were two or more siblings with severe mental retardation (Benson *et al.*, 1970 in 'Errors of Phenylalanine, Thyroxine and Testosterone Metabolism', p. 45, eds W. Hamilton and F. P. Hudson, E and S. Livingstone, Edinburgh) and found three families in which two or more siblings had similar defects of renal tubular reabsorption. In one of these (three brothers and one sister) there was hydroxylysinuria (Benson *et al.*, 1969, *Arch. Dis. Child.* **44**, 134). In this family, all the retarded siblings had apparently normal development in the first year of life. After this age there was progressive deterioration of cerebral function characterised by hyperkinesis, major fits, attacks of trembling and myoclonic spasms. Parker *et al.*, 1970 (*Lancet* **1**, 1119) recently described another family with hydroxylysinuria. The second family in our study with defects of renal tubular reabsorption consisted of two sisters who had β-alaninuria. They differed from the child described by Scriver *et al.*, 1966 (*New Eng. J. Med.* **274**, 635) since our patients had no hyper-β-alaninaemia. The third family consisted of two brothers who persistently had a high urinary excretion of 1-methylhistidine.

Polani: Was there a difference in the sex incidence in patients with imino glycinuria?

Scriver: No difference has been observed.

Komrower: In newborn infants it seems that a most striking indication of immaturity of the renal system is lack of glycine reabsorption.

Scriver: We wondered if deficiency of microvilli contributed to immaturity of amino acid renal tubular reabsorption in the newborn. We believe that differentiation of specific activity of transfer sites is as important, if not more so, than insufficiency of absorptive area in the tubule during early life.

Hartnup Disease

JOHN B. JEPSON

Courtauld Institute of Biochemistry,
The Middlesex Hospital Medical School,
London, W1P 5PR

In 1951, a Mrs. Hartnup called at the Middlesex Hospital, London, with her son Edward, then aged 12. She declared that he had pellagra–a startling do-it-yourself diagnosis. It soon transpired that her older daughter Pat had exactly the same symptoms in 1937–a red scaly rash on exposed parts, cerebellar ataxia, and occasional episodes of mental instability–and at that time the Middlesex Hospital had decided that pellagra was the only condition which could fit the situation. This was the beginning of a vast and continuing series of metabolic investigations of what proved to be a 'new' genetic disease; and yet not a disease in the standard sense, because it may cause only the mildest inconvenience–Hartnup Disorder would be better, with Hartnup Peculiarity better still. Fully documented accounts of the condition, including biochemical, clinical and genetical aspects, are available (Jepson 1966; Milne 1969). We consider it at this Study Group because, under special circumstances, the disorder may lead to temporary derangement of mental, psychological and neurological status. Whether it is appropriate to the session entitled 'Renal Tubular Function' is another matter, because the renal defect of Hartnup Disorder is clinically its least important aspect.

Dietary pellagra, caused by a dietary deficiency of nicotinic acid and nicotinamide, was obviously untenable in the Hartnup children though acute episodes were precipitated in them, as in later cases, by dietary stress. Fortunately, between the two hospital admissions in 1937 and 1951, C. E. Dent had demonstrated the fascinating possibilities of urinary amino acid chromatography; in lieu of anything else more rational to do, we applied this to a random urine sample from Edward Hartnup, and found an extraordinary amino acid excretion pattern, which was later recognised as showing faulty renal reabsorption of the monoamino monocarboxylic acids, but not those of the group characterising cystinuria or glycine-iminoaciduria (Fig. 1). Four of the eight children of the Hartnup family showed this characteristic anomaly (Fig. 2) and this has been the only sure diagnostic test for the few dozen subjects found subsequently. The renal clearances for the amino acids involved are very high—that for histidine approaches the glomerular filtration rate; in general, the renal reabsorption of the affected amino acids is 50-80%, against a normal value of 98% (Scriver, 1969). Scriver and Raine, earlier in this session, have described our present view of the several mechanisms, with different capacities and specificities, which

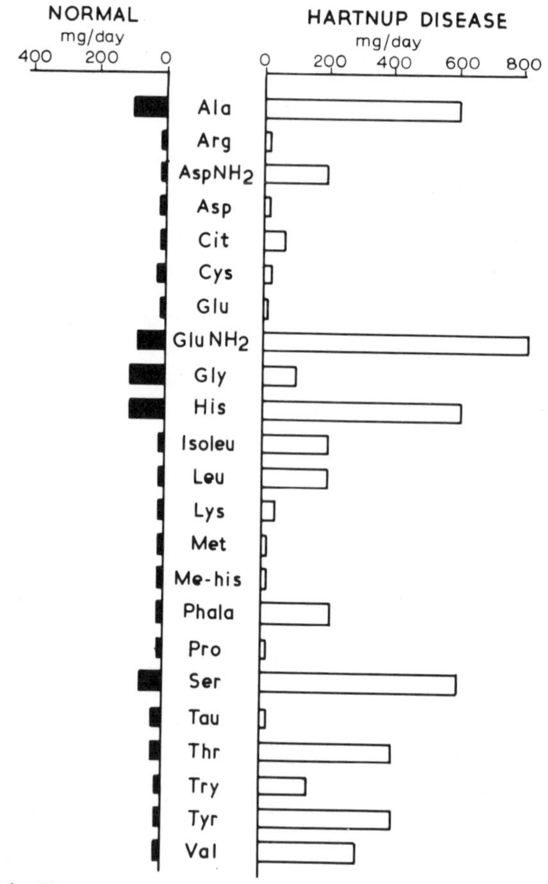

Fig. 1. The excretion of free amino acids in the urine of normal subjects and Hartnup patients.

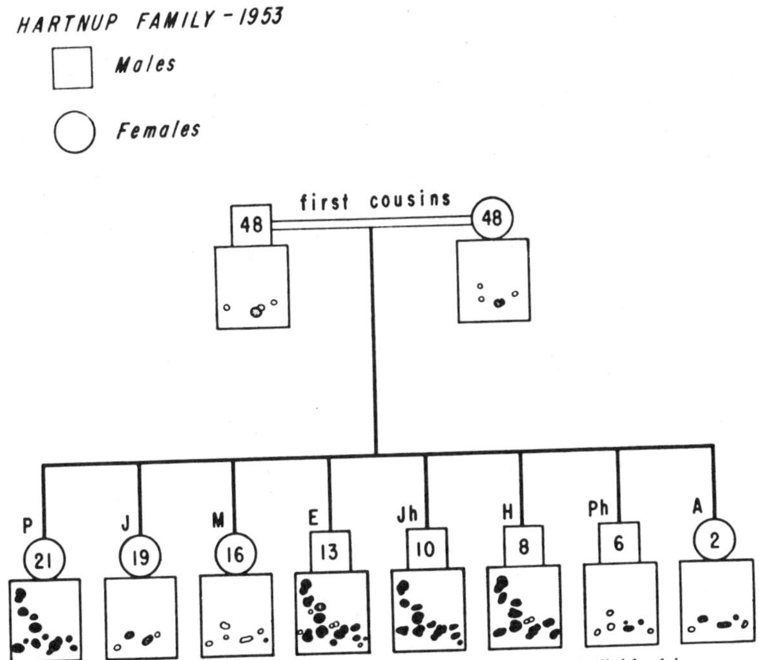

Fig. 2. The original Hartnup family. The age of each individual in 1953 is given in years.

Each square shows the two-dimensional chromatographic pattern of that individual's urinary amino acids.

operate for the active transport of free amino acids across membranes.

One of the Hartnup girls, M. H., was mentally subnormal, but she did not manifest the renal disorder. However, it *appeared* that the intelligence of an affected child was lower than that of the other affected sibs who were younger, hence the original suggestion that the disorder caused mental deterioration with age (Baron *et al.*, 1956). Later cases have shown that this is not tenable; if anything, the opposite is more usual and there is improvement in all aspects of the condition as the stresses of growth and puberty subside.

Something over forty cases have now been reported and this is a very small number from which to generalise.

Almost all of these affected subjects had developed a photosensitive rash before the age of 10; half of them have, or had, reversible neurological or neuropsychiatric symptoms—ataxia, sudden fainting, hallucinations, emotional instability, complete delirium, a range similar to that with pellagra. Only nine are described as mentally retarded; only one is mentally defective, and she is a deaf-mute discovered in a mental hospital through routine screening by urinary chromatography. The family tree which includes this subject, A, is shown in Fig. 3, and full details are given in the report by Pomeroy *et al.* (1968). This is the only American family fully documented and includes the oldest known Hartnup patient, A, and the only patients, B and C, with offspring.

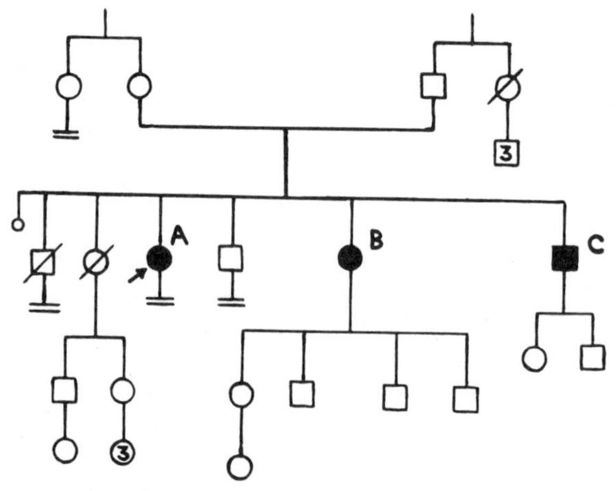

Fig. 3. Pedigree of the individuals with Hartnup Disorder reported by Pomeroy *et al.* (1968).

Apart from the sibs A, B and C, none of the individuals included in Fig. 3 show the characteristic amino acid excretion. Apart from A, all are in good health, mentally well, and of normal-to-high intelligence.

The pattern of inheritance is that expected for an autosomal recessive characteristic, manifest in the 'double dose' condition only. There is no biochemical test which identifies the heterozygote, though there are indications that obligate heterozygotes (e.g. the children of C) have learnt by experience to avoid exposure to sunlight. The only family which does not unequivocally fit this mode of inheritance is the extended Japanese family reported by Oyanagi *et al.* (1967), with the pedigree shown in Fig. 4. Two of the four living children from the second-cousin marriage had the biochemical *and* severe clinical characteristics of Hartnup disorder. The other two children were symptomless, but were described as showing the Hartnup excretion patterns, as also did the healthy grandfathers and two first cousins. Oyanagi called these 'carrier members of Hartnup disease' and this conclusion fits recessive transmission within the pedigree perfectly well, were it not for the fact that no other obligate heterozygote shows the slightest defect in renal amino acid reabsorption. The two obvious cases were very severely affected, neurologically and psychologically, so the allele concerned could be a vicious variant that even shows biochemically in single dose, but then the parents should be affected. Alternatively, the biochemically-affected individuals could all be

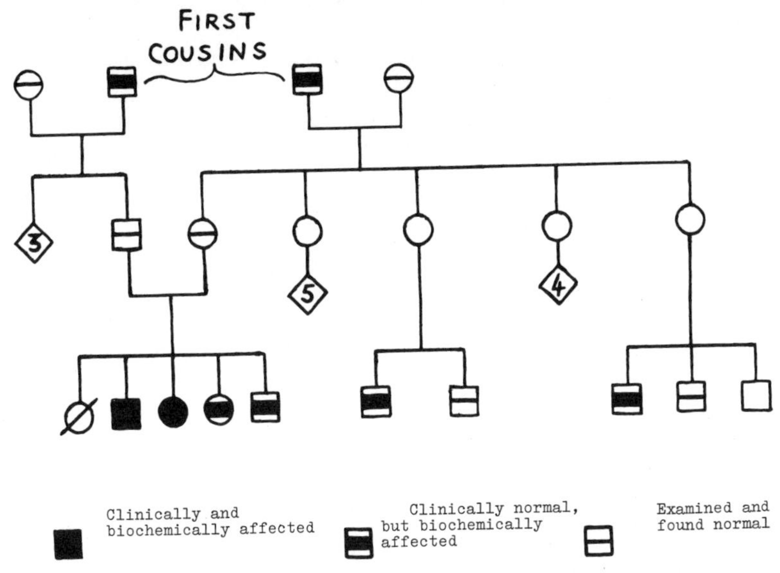

Fig. 4. Pedigree of the individuals with Hartnup Disorder reported by Oyanagi *et al.* (1967).

homozygotes, due to the introduction of apparently-unrelated heterozygotes into the family, but not shown in Fig. 4.

The renal defect leads to some loss of amino acids, including tryptophan and other essential amino acids, but this loss is not large enough to be of much nutritional significance. However, there is also a large but variable urinary excretion of tryptophan degradation products–indoxyl sulphate (indican), indolylacetic acid and so on–especially following an oral load of free tryptophan. This excretion of tryptophan metabolites is abolished if a Hartnup patient has his gut sterilised with antibiotics, a procedure which does not effect the specific excretion of free amino acids (including tryptophan). This led Milne *et al.* (1961) to postulate that the specific defect for amino acid absorption applied not only in the renal tubule but also in the gut. Tryptophan absorption would be delayed, and the amino acid would therefore be subjected to bacterial metabolism in the lower bowel; the acidic products of this attack would be absorbed, and excreted in the urine. Direct measurement of free serum tryptophan following oral or intravenous tryptophan loads confirms the intestinal defect (Wong and Pillai, 1966).

The time course for the urinary excretion of indican and indolylacetic acid following oral tryptophan (Fig. 5) demonstrates the intestinal retention of the free amino acid.

Thus, Hartnup disorder is characterised by diminished gut absorption of free tryptophan, increased gut destruction of tryptophan, and increased renal loss of tryptophan, all of which amounts to the unavailability of a substantial quantity of the dietary intake of this

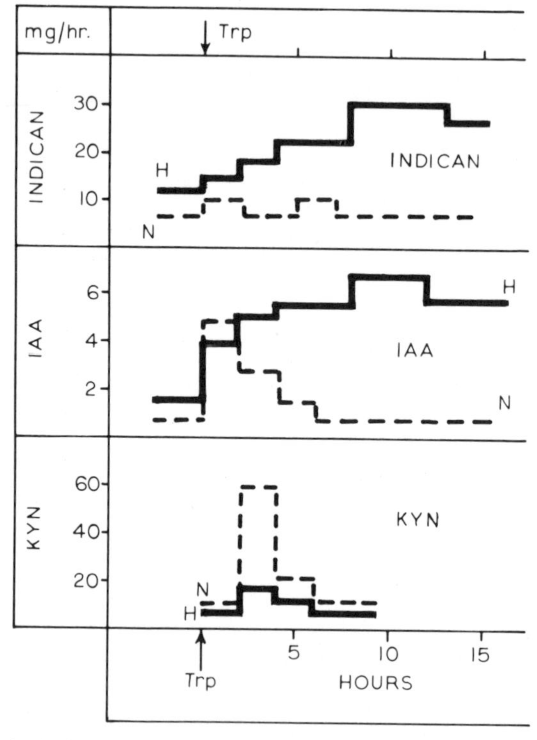

Fig. 5. Urinary excretion (mg/hour) of tryptophan derivatives following oral L-tryptophan (about 70 mg/kg body wt) administered to normal subjects (N – – –) and to Hartnup patients (H—). These are 'idealised' responses, averaged from several investigations. [IAA = indolylacetic acid; KYN = kynurenine].

essential amino acid. This correlates well with the clinical similarity of Hartnup Disorder to pellagra, because one of the metabolic pathways for tryptophan utilisation leads to nicotinamide via kynurenine. The small amount of kynurenine excreted by Hartnup patients following oral tryptophan (Fig. 5), confirms that these individuals are often walking close to the edge of nicotinamide insufficiency, because tryptophan is not always available for the production of endogenous nicotinamide to supplement dietary nicotinamide, as in normal humans. Clinically acute cases of Hartnup disease always come from economically-deprived families or under the stress of temporary malnutrition.

The recent case described by Navab and Asatoor (1970) involved a vegetarian girl of 25 who precipitated her first attack by visiting Kenya where maize had to constitute most of her diet. The grossly abnormal mental state which developed improved dramatically within 24 hours of starting therapy with intravenous nicotinamide.

It is a distinct possibility that the nicotinamide required by particular tissues or cells, or in specific metabolic situations, has to be synthesised *in situ* from tryptophan, because ready made (dietary) nicotinamide is not an adequate substitute. Nevertheless, excess dietary nicotinamide would spare tryptophan for this obligatory 'nicotinogenic' role. There is evidence that free nicotinamide is not incorporated into rat brain cells, but, conversely, rat brain appears not to possess all the enzymes required for the tryptophan-to-nicotinamide conversion. Because dietary pellagra can cause permanent mental deficiency, it seems important to search for a special function of nicotinamide in relation to mental health and disease, apart from its obvious role as a coenzyme component in cerebral enzymic systems.

Since the discovery of Hartnup disease, there have been several suggestions that an endogenous 'toxic' agent is responsible for some or all of the neuropsychiatric manifestations. Tryptamine is an obvious candidate, but recent work by Miss J. Dayman in my laboratory has shown that blood tryptamine is hardly detectable, and that Hartnup patients show *less* of a rise in urinary tryptamine after oral tryptophan loading than do normals. Even a Hartnup patient liberally dosed with ethanol showed no detectable tryptophol accumulation, which would have been expected if accumulating tryptamine was being metabolised through the corresponding aldehyde (Davis *et al.*, 1967). The rapid improvement brought about by nicotinamide therapy (in the neurological symptoms if not in the rash) probably disposes of the toxic agent theory, unless nicotinamide is required for the toxic agent's disposal!

As already discussed, oral amino acid tolerance measurements show a greatly diminished gut absorption of free tryptophan in Hartnup patients, and the same is true of phenylalanine and histidine. The question arises as to how Hartnup patients manage to obtain adequate supplies of these essential amino acids. A clue has been provided by Milne and his colleagues (Navab and Asatoor, 1970; Asatoor, 1970). They compared the absorption into serum of amino acids given orally in the form of dipeptides with the absorption of the same two constituent amino acids mixed together in the free form. For example, an oral tolerance test with glycyl-tryptophan was compared with the tolerance test using the equivalent amount of mixed glycine and tryptophan, the serum tryptophan level being measured at intervals over 3 hours. In normal people, as expected, free tryptophan was absorbed faster and more completely than tryptophan from the dipeptide glycyl-tryptophan. In a Hartnup patient, the situation was reversed (Fig. 6). Other dipeptides, involving phenylalanine or histidine, show an identical effect. The suggestion is that the human intestine uses sites or mechanisms for the absorption of dipeptides (and possibly higher peptides) which are distinct from those for free amino acid absorption. Only the latter are defective in Hartnup disease, and then only for the one group of amino acids, so that Hartnup patients can make up by peptide absorption for some of the diminished capacity of

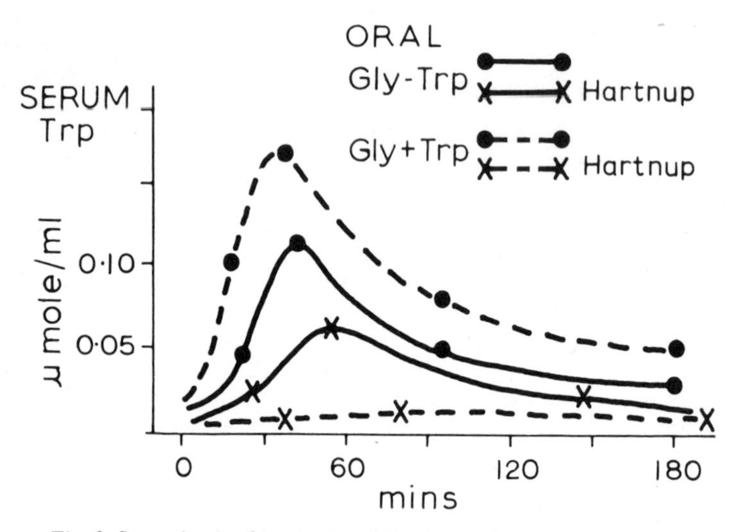

Fig. 6. Serum levels of tryptophan following equivalent oral loads of glycyl-tryptophan (——) or glycine-plus-tryptophan (– – –) in normal subjects (●●) and in a Hartnup patient (xx) (Asatoor *et al.*, 1970).

the absorption mechanisms for free amino acids. Of course, these tolerance experiments use oral doses which are large by comparison with the amounts encountered during the normal processes of digestion, and several systems with quite different capabilities may be involved under physiological conditions. By analogy with renal tissue, Hartnup patients will have their aberration in the shared, high-capacity system for intestinal absorption, while the low-capacity systems for specific amino acids still operate.

REFERENCES

Asatoor, A. M., Cheng, B., Edwards, K. D. G., Lant, A. F., Matthews, D. M., Milne, M. D., Navab, F. and Richards, A. J. (1970). Intestinal absorption of two peptides in Hartnup disease. *Gut* **11,** 380.

Baron, D. N., Dent, C. E., Harris, H., Hart, E. W. and Jepson, J. B. (1956). Hereditary pellagra-like skin rash with temporary cerebellar ataxia, constant renal aminoaciduria and other bizarre biochemical features. *Lancet* **2,** 121.

Davis, V. E., Brown, H., Huff, J. A. and Cashaw, J. L. (1967). Alteration of serotonin metabolism to 5-hydroxytryptophol by ethanol ingestion in man. *J. Lab. Clin. Med.* **69,** 132.

Jepson, J. B. (1966). In 'Metabolic Basis of Inherited Disease', 2nd edition, (ed. J. Stanbury), Chapter 57. McGraw-Hill.

Milne, M. D. (1969). Hartnup disease. *Biochem. J.* **111,** 3P.

Milne, M. D., Asatoor, A. and Loughridge, L. (1961). Hartnup disease and cystinuria. *Lancet* **1,** 51.

Navab, F. and Asatoor, A. M. (1970). Studies on intestinal absorption of amino acids and a dipeptide in a case of Hartnup disease. *Gut* **11,** 373.

Oyanagi, K., Takagi, M., Kitabatake, M. and Nakao, T. (1967). Hartnup disease. *Tohoku J. exp. Med.* **91,** 383.

Pomeroy, J., Efron, M. L., Dayman, J. and Hoefnagel, D. (1968). Hartnup disorder in a New England family. *New Engl. J. Med.* **278,** 1214.

Scriver, C. R. (1969). The human biochemical genetics of amino acid transport. *Pediatrics* **44,** 348.

Wong, P. W. K. and Pillai, P. M. (1966). Clinical and biochemical observations in two cases of Hartnup disease. *Arch. dis. Childh.* **41,** 383.

DISCUSSION

Milne: I should like to ask Dr. Scriver whether he studied peptide absorption in the microoorganisms with amino acid transport defects.

Scriver: Yes, we looked at the specificity of amino acid as distinct from peptide absorption sites and were able to confirm that they were distinct.

I was very interested in Professor Milne's explanation on how subjects with Hartnup syndrome absorb amino acids as peptides. In order for this mechanism to be effective a substantial proportion of dietary protein must be digested to dipeptides only. Is there any information on whether this is the case? I should be interested to know whether the effect of dipeptide concentration on absorption rates was studied.

The absence of excessive methionininuria in Hartnup syndrome may be simply that the renal threshold for methionine on the low Km system is not reached. We found a comparable situation in a family with iminoglycinaemia trait members who had no hyperprolinuria. The explanation was that the plasma proline concentration did not exceed the renal threshold.

Milne: To answer Dr. Scriver's point only about 10% of the products of protein digestion in the jejunum are in the form of free amino acids, the remaining 90% are in the form of oligopeptides.

There is some data which tends to support the dipeptide absorption explanation rather than dual site hypothesis suggested by Dr. Scriver. In subjects with Hartnup syndrome there is very little rise in plasma levels following administration of some free amino acids, such as tryptophan, histidine and phenylalanine. If Dr. Scriver's theory was correct and a continuing low level of absorption of amino acids occurred sufficiently to maintain life, then one would expect to be able to demonstrate this. On the other hand in patients with Hartnup syndrome considerable absorption of dipeptides can be demonstrated.

Dr. Matthews at Westminster Hospital and later myself and others have shown that in normal individuals absorption of glycine is greater when it is in the form of dipeptides (glycylglycine) than as tripeptides (glycylglycylglycine) and greater as tripeptides than as tetrapeptides (glycylglycylglycylglycine).

Scriver: I agree that loading tests with tryptophan and other free amino acids indicate that there is an absorption defect. The large amounts used in such tests, e.g. about 2 g of tryptophan, or the nutritional requirements for about 2 days, would exceed the physiological capacity of the low Km system for tryptophan reabsorptiom by several 100-fold and the data suggest a defect of high Km absorption sites. Such heavy loading tests could not demonstrate any defect from low Km sites which would be functional only under normal physiological conditions but would continue to absorb low concentrations of tryptophan over long periods of time and might therefore be effective in maintaining adequate nutrition.

Milne: Yes, the present data do not exclude this. It is possible that in Hartnup syndrome nutrition is maintained both by absorption of amino acids at low Km sites and of dipeptides.

Yielding: I think that a study of dipeptide absorption would be an excellent way to study the nature of the recognition sites for absorption. One could compare the absorption of glycyltryptophan with tryptophanylglycine.

Jepson: One reason that people did not suspect that dipeptide absorption could be important is that dipeptides can hardly be demonstrated in the plasma.

Milne: I myself think that the reason for the low peptide content of blood and urine is probably because of the very high peptidase activity in many tissues including kidney. However, some types of peptides, for example γ-glutamyl peptides, and β-aspartyl peptides probably appear in urine because they cannot be hydrolysed owing to the lack of the appropriate intracellular peptidases. We have compared the intestinal absorption of peptides to that of free amino acids in cystinurics and have shown that they can absorb lysylglycine much better than they can absorb free glycine or lysine.

Raine: How does Dr. Scriver explain the excessive urinary excretion of cystine and lysine in cystinurics since these two amino acids do not share the same system for renal membrane transport.

Scriver: One may postulate that in cystinuria, only ornithine, arginine and lysine share an influx system in kidney and that there is instead an efflux abnormality in cystinuria, involving a system shared by cysteine and the three dibasics (the Segal hypothesis). The presence of systems with different specificities in different tissues or on different sides of the membrane is not a novel proposal. It is known for glucose, for example.

Milne: I find it very difficult to accept that the intestinal transport defect in cystinuria is an abnormality of efflux. Evidence points strongly that it is influx which is abnormal. There is also some evidence that there may be two systems for intestinal absorption of glucose. In the condition of glucose-galactose malabsorption the ratios of glucose and galactose which are absorbed during *in vivo* studies are radically different to the normal. Thus in the normal state galactose absorption is greater than that of glucose while in galactose-glucose malabsorption glucose absorption is about twice as fast as galactose. This seems to indicate the presence of at least two systems for glucose absorption.

Bradford: I should like to ask two questions: first, to what extent can mental retardation and neurological signs be attributed to nicotinamide deficiency and second, has a Hartnup-like disorder been detected in laboratory animals?

Jepson: Although there is clearly some connection between some aspects of the clinical features of Hartnup's syndrome and nicotinamide deficiency, for example, the pellagra-like skin, the neurological symptoms are not wholly characteristic of those found in pellagra.

No equivalent situation has yet been found in animals. An

interesting abnormality of tryptophan *metabolism* was recently described in humans. This consisted of a block in the tryptophan-kynurenine pathway which resulted in the accumulation of tryptophan. Clinically there was a pellagra-like rash and some neurological abnormalities but not the complete picture of Hartnup syndrome (Tada *et al.*, 1963, *Tohoku J. Exp. Med.* **80**, 118).

Polani: Has anyone searched for genetic pre-disposition as a cause for genuine pellagra?

Allison: Not as far as I know. One should remember that pellagrins often have multiple nutritional deficiencies, for example, of protein and several vitamins besides nicotinamide.

Jepson: This is emphasised by a report (Hankes *et al.*, 1970, *Fed. Proc.* 3628) which states that administration of tryptophan to genuine pellagrins produced very high blood levels of kynurenic acid and other kynurenine derivatives. This was attributed to multiple interferences with the tryptophan-nicotinamide pathway including Vitamin B6 deficiency.

Komrower: Some time ago we described a child with hydroxy-kynureninuria and mental retardation who on one occasion developed a photosensitivity rash but did not have ataxia. We attributed the disorder to a block in the tryptophan-nicotinamide pathway. The development of the rash in spite of ample dietary nicotinamide supports Professor Jepson's comment that nicotinamide synthesised intracellularly appears to be much more effective in preventing pellagra, than preformed nicotinamide in the diet.

Jepson: An interesting feature of the rash of patients with Hartnup syndrome is that one does not seem able to produce it to order. We have failed to provoke a rash in these subjects even after exposing them to ultra-violet radiation.

Milne: In the family with Hartnup syndrome which I studied the rash would appear when the patient was on a pellagra-producing diet but this would happen much sooner than in normal subjects on such a diet.

It has recently been reported that Indians develop pellagra much more easily when on a diet of millet than on a diet of maize. This was attributed to the higher leucine content of millet than maize. Pellagra could be precipitated in the maize eaters who had been given leucine supplements. This effect seemed to be specific for leucine and could not be produced by addition of isoleucine to the diet.

Scriver: I wonder whether this is due to the fact that the Km for leucine reabsorption is very low?

Session III

Lysosomal Function and Disease

Lysosomes in Nervous Tissue

J. T. DINGLE

Tissue Physiology Department
Strangeways Research Laboratory, Cambridge

Lysosomes are known to play an important role in the physiology and pathology of almost all cells and tissues. The diverse functions of these organelles have recently been comprehensively reviewed by de Duve and others (Dingle and Fell, 1969) and it would not be possible to attempt to summarise the wealth of information that is now available. In the context of this symposium it would seem most useful to devote this chapter to a short review of the recent work on the role of lysosomes in nervous tissues and then to outline briefly some unpublished studies in which immunoenzymic methods have been developed to investigate the role of lysosomes in connective tissues. These methods would appear to be applicable to the further study of lysosomes in nervous tissues.

Methods Used for Study of Lysosomes in Nervous Tissues

Light Microscopy

Lysosomes are usually identified by the cytochemical staining of typical acid hydrolases (for review see Gahan 1967); acid phosphatase and arylsulphatase, two of the most useful markers for lysosomes, have been demonstrated in the perikaryon region of normal neurons. In the neurons of rat dorsal root and nodosal ganglia, larger structures, probably associated with the Golgi apparatus also stain for enzyme activity (see Holtzmann 1969 for review). Fewer lysosomes are present in the more distal regions of axons.

Lysosomes in neural tissues may also be demonstrated by selective staining with certain cationic dyes, such as acridine orange or toluidine blue (Koenig, 1968). Neural lysosomes, like those in other tissues, emit a natural yellow fluoresence upon activation at 330 mμ (Koenig, 1963).

Electron Microscopy

Numerous dense bodies are present in the perikarya of normal neurons. These bodies have a compact electron dense matrix which has been shown to stain for acid phosphatase. According to Koenig (1969) such enzyme-positive bodies are more numerous in grey matter than in white matter, intra-axonal lysosomes being rare in both white matter and peripheral nerve. The perikaryon is also rich in multivesicular bodies, which have been shown to contain acid phosphatase activity, particularly in conditions of increased enzyme synthesis (Holtzman and Novikoff, 1965). Arylsulphatase has also been demonstrated in both multivesicular bodies and dense bodies of neurons (Holtzman, 1969), whilst Torach and Barrnett (1962) have reported the presence of esterase. Autophagic vacuoles appear to be relatively infrequent in normal neurons.

Novikoff (1967) has pointed out the close relationship between lysosomes and the Golgi apparatus in neurons. He considers that the agranular membranes associated with the inner concave surface of the apparatus are involved in the packaging of acid hydrolases formed in the endoplasmic reticulum. He has termed this region GERL; it is probable that it has a similar function in other tissues also.

Neurons contain dense bodies which are rich in lipofuscin. There appears to be a correlation between the appearance of these bodies and the ageing process, but vitamin D deficiency and other degenerative changes may also give rise to an increased number of these organelles. The mode of formation is unknown.

Biochemical Studies

Koenig (1969) has pointed out that although lysosome-rich fractions may be readily obtained from brain homogenates, the subcellular distribution pattern of lysosomal enzymes in brain homogenates is very sensitive to the fractionation procedure. Sellinger and Hiatt (1968) found evidence for two distinct populations of lysosomes in rat brain, similar findings have been reported by other authors. There seems little doubt that the largest population of lysosomes are rather more dense than mitochondria, the presence of some of these dense lysosomes inside fragments of nerve endings or axons may contribute to some of the difficulties that have been experienced in obtaining reproducible sedimentation patterns. A variety of acid hydrolases including cathepsin, galactosidase, nucleases, phosphatase, glucuronidase and arylsulphatase have been demonstrated in these particles. The enzymes have been shown to be released by thermal activation, detergents, low pH and the other methods which are now known to disrupt lysosomal membrane integrity.

These microscopic and biochemical studies show beyond doubt that the cells of nervous tissues contain lysosomes and that their properties are not very dissimilar from those of other tissues. However, much less is known about the physiological and pathological role of these lysosomes in nervous tissues.

Lysosomes in Injured Nervous Tissue

Holtzman and Novikoff (1965) noted focal accumulations of acid phosphatase, distal to the injury, in axons undergoing changes subsequent to mechanical injury, whilst several authors have reported that large acid phosphatase positive organelles increase in number in neurons that have been subjected to mechanical injury, radiation injury or anoxia. The number of autophagic vacuoles in the perikarya of injured neurons increases considerably, whilst many large bodies containing lamellae and other contents, which may have been derived from autophagic vacuoles, are also present. Bodies derived from autophagic vacuoles are also present in the proximal segment of interrupted nerves, where they may be larger than mitochondria (Holtzman, 1969). In the distal segment of interrupted nerves, most of these bodies are smaller. In view of the possibility of artefacts and because of the lack of cytochemical information Holtzman recommends caution in identifying structures in degenerating axons as lysosomes.

Various poisons cause change in neuronal lysosomes. Thus fluorocitrate causes an apparent movement of dense bodies from spinal neurons into axons, leading to the formation of axonal ballons. (Koenig, 1969). It would appear that some axons may also contain free acid phosphatase activity after fluorocitrate poisoning, though again the possibility of artefacts must be considered in any situation in which diffuse staining occurs. Many other agents including neutral red, actinomycin D, diphtheria toxin and tetrazolium salts cause changes in neural lysosomes, principally by an increase in the number of autophagic vacuoles.

Lysosomes in Neural Pathology

Relatively little information is available about the role of lysosomes in diseases affecting neurons, most biochemical studies have been confined to brain tumours. Thus β-glucuronidase has been shown to be increased in both tumour tissue (Allen, 1961) and in the cerebrospinal fluid of tumour cases (Allen and Reagen, 1964). Increased levels of lysosomal enzymes are also found in the cerebrospinal fluid in acute demyelinating disease, in diabetic neuropathy, Barré syndrome and in meningeal carcinomatosis. These findings are consistent with an extracellular secretion of lysosomal enzymes, which also has been found to occur in other tissues under pathological conditions (Dingle, 1969).

The storage diseases, lipidoses and polysaccharidoses, have been studied in some detail. In Tay-Sachs disease (Wallace *et al.*, 1964) numerous large inclusion bodies are present in the perikarya, these bodies contain acid hydrolases and may be considered to be lysosomes. These organelles also contain considerable amounts of gangliosides and similar inclusion bodies have been observed in infantile lipidoses (Gonatas and Gonatas, 1965), Niemann-Pick disease, and Jakob-Creutzfeld disease. In Alzheimer's disease acid phosphatase positive dense bodies have been shown to accumulate in both the axon and the synaptic endings (Suzuki and Terry, 1967). Holtzman (1969) considers that in many of these conditions the changes in lysosomal morphology is a secondary effect of the disorder, a conclusion that is substantiated by much other work on a variety of tissues, in many of which it has been found that the lysosomal system is very susceptible to change in the metabolic activity of the cell.

The problem of assessing the role of lysosomal system and of endeavouring to establish which individual enzymes are concerned in a pathological lesion has been studied by members of the Tissue Physiology Department of the Strangeways Laboratory for several years. We have recently developed specific immunoenzymic methods which enable us to study the synthesis, translocation and specific function of individual lysosomal enzymes. The remainder of this chapter will be devoted to a short review of this work.

Immunoenzymology of Lysosomal Enzyme

The lysosomal system plays a key role in the catabolism of the proteoglycans of connective tissues and there is strong, but circumstantial evidence that this process is mediated principally by lysosomal cathepsin D. Other workers have proposed the involvement of hyaluronidase, cathepsin B_1 and β-xylosidase. If the hypothesis that cathepsin D is the enzyme primarily responsible for the breakdown of cartilage matrix proves to be correct, it has important implications in the pathology of connective tissue disease, particularly arthritis (see Dingle 1970 for review). This hypothesis could be tested by the specific inhibition of the enzyme in a biological test system; since no suitable chemical inhibitor is known, my colleagues and I have examined the possibility of using antibodies to cathepsin D for this purpose (Barrett *et al.*, 1970).

An essential first step was the complete purification of the enzyme from the three species (human, rabbit and chicken) in which we are interested. Barrett (1970) has developed a scheme of purification which enables homogeneous enzyme protein to be prepared from the livers of all three species. The method involves fractionation with acetone, column chromatography on DEAE-cellulose and CM-cellulose and preparative isoelectric focusing. Three main forms, or isoenzymes, are obtained in a pure state from chicken and human liver; the forms of cathepsin D in rabbit liver are more numerous and as yet only small quantities of any one isoenzyme have been completely purified. The isoenzymes do not appear to differ in their molecular weight (45,000) or pH optimum, and they give reactions of immunology identity. Barrett has found that isoelectric focusing of crude tissue extracts shows patterns of isoenzymes very similar to that obtained in the final purification.

The possession of pure cathepsin D has enabled us to raise antisera to it in rabbits and sheep; these antisera were specific inhibitors of the enzyme in the range pH 5-7; the action of the enzyme on haemoglobin at pH 5.0 was completely inhibited by 5-10 μl of antiserum/unit enzyme and pure anti-cathepsin D antibodies prepared on an immunoadsorbent column were also effective. Experiments with this material demonstrated that the antigen-antibody complex was precipitated when 2-4 antibody molecules combined with one molecule of cathepsin D; this complex was still enzymically active, however, whereas all activity was lost when 6-7 molecules of antibody were bound to each enzyme molecule. No precipitation occurred when univalent antibody fragments prepared from a pepsin digest of antiserum IgG were reacted with cathepsin D, but this material was as potent an inhibitor as the undigested material, indicating that precipitation is not essential for inhibition. The antisera had no effect on the activity of various other lysosomal enzymes.

The antisera inhibited the action of pure cathepsin D on purified cartilage proteoglycan and on enzyme-free [^{35}S] labelled cartilage. As a model system in which to study the breakdown of articular tissues, we have used the autolytic degradation of cartilage matrix in the range pH 5-7. The release of the polysaccharide component of the protein-polysaccharide complex was measured by chemical and radiochemical means, and the process was found to be 95% inhibited by anti- (cathepsin D) antisera; we take this as direct evidence for an important role of cathepsin D in this type of cartilage breakdown (Weston *et al.*, 1969). It remains to be seen what part the enzyme plays in living tissues, in different species and in remodelling and pathologically reabsorbing tissues. The recent preparation of antisera to other tissue proteases of human and animal origin should give much new information on the role of these enzymes in connective tissue pathology, particularly arthritis. Specific antisera to selected lysosomal enzymes are being used to localise these enzymes at the light and electron microscope levels after tagging with suitable markers. Experiments are also in progress to determine the relationship of lysosomal enzyme synthesis to secretion; the method makes use of the specific precipitation of pulse labelled enzyme by antibody.

Conclusions

Immunoenzymic methods are making a significant contribution to an understanding of the role of the lysosomal system in the metabolism of connective tissues. Many of these methods could readily be applied to the study of lysosomal function in nervous tissues and could do much to further our, as yet, slight knowledge of normal lysosomal function in axon endings. In particular antibodies to the lysosomal proteases and cerebroside galactosidase might yield much needed information on the mechanism of turnover of axoplasmic constituents. Studies on the specific synthesis and location of lysosomal enzyme in storage diseases, which are now capable of being carried out by immunoenzymic methods, would also be likely to yield valuable information.

REFERENCES

Allen, N. (1961). *Neurology* **11,** 578.
Allen, N. and Reagen, E. (1964). *Arch. Neurol.* **11,** 144.
Barrett, A. J. (1970). *Biochem. J.* **117,** 601.
Barrett, A. J., Weston, P. D. and Dingle, J. T. (1970). *Biochem. J.* in press.
Dingel, J. T. (1969). In 'Lysosomes in Biology and Pathology', (eds J. T. Dingle and H. B. Fell), Vol. II, p. 421. Amsterdam: North Holland.
Dingle, J. T. (1970). In 'Tissue Proteinases', (eds A. J. Barrett and J. T. Dingle). Amsterdam: North Holland (in press).
Dingle, J. T. and Fell, H. B. (eds) (1969). 'Lysosomes in Biology and Pathology', Vols. I and II. Amsterdam: North Holland.
Gahan, P. B. (1967). *Intern. Rev. Cytol.* **21,** 1.
Gonatas, N. K. and Gonatas, J. (1965). *J. Neuropathol.* **26,** 25.
Holtzman, E. (1969). In 'Lysosomes in Biology and Pathology' (eds J. T. Dingle and H. B. Fell), Vol. I p. 192. Amsterdam: North Holland.
Holtzman, E. and Novikoff, A. B. (1965). *J. Cell Biol.* **27,** 651.
Koenig, H. (1963). *J. Histochem. Cytochem.* **11,** 120.
Koenig, H. (1968). In 'Barrier Systems in the Brain' (eds A. Lajtha and D. M. Ford), p. 87. Amsterdam: Elsevier.
Koenig, H. (1969). *Science* **164,** 310.
Novikoff, (1967). In 'Enzyme localisation and ultrastructure of neurons and lysosomes in nerve cells' (ed. H. Hyden), p. 255. Elsevier Pub. Co.

Sellinger, O. Z. and Hiatt, R. A. (1968). *Brain Res.* **7**, 191.
Suzuki, K. and Terry, R. D. (1967). *Acta Neuropathol.* **8**, 276.
Torach, R. M. and Barrnett, R. J. (1962). *Exptl Neurol.* **6**, 224.
Wallace, B. J., Avonson, S. M. and Volk, B. W. (1964). *J. Neuropathol.* **25**, 76.
Weston, P. D., Barrett, A. J. and Dingle, J. T. (1969). *Nature, Lond.* **222**, 285.

DISCUSSION

McIlwain: Have you demonstrated lysosomal enzyme activity in the regions of axon terminals and the spine processes where one would expect maximum degradation of protein to take place?
Dingle: With the substrates normally used, cytochemical staining for lysosomal enzymes in axons is only slight, but it is possible, though not likely that, lysosomal enzyme activity might be demonstrated in axons using other substrates.
Allison: Holtzman gave a paper at the Gordon Conference on lysosomes this year. He stressed that there were large numbers of lysosomes in neurones, only a few in the axon near the neurone but a large number at the other end of the axon, and these may very well be involved in degrading material which passes down the axon.
Bradford: The movement of particles in axons was shown very clearly in the study by Pomerat (Dynamic Aspects of the Neurone on Tissue Culture, 1964, 16 mm sound available from Pasadena Institute for Medical Research U.S.A.) of neurones in culture using time-lapse cinematography. Movement of particles was mainly away from the neurones but in some axons was clearly bidirectional.
Yielding: Is it thought possible that permease-like lysosomal enzymes could act in the reverse direction and synthesise peptide bonds? The free energy considerations would oppose this in aqueous media but conditions might favour peptide bond synthesis in non-aqueous, apolar media.
Dingle: I would have thought this most unlikely in digestive vacuoles and I don't know whether it would occur in GERL vesicles. This could be tested experimentally.
Polani: I wonder whether this type of protein synthesis might occur in the region of the Golgi apparatus of spermatids?
Allison: It is perfectly true that the acrosome is formed along the Golgi apparatus but I think it is very unlikely that it is concerned with protein synthesis.
McIlwain: I think the lack of genetic direction is a serious objection to the idea that protein synthesis might occur in lysosomes.

Yielding: I completely accept that the sequences of amino acids in peptides are coded by genetic mechanisms. I was thinking that preformed polypeptides for example in the form of repeating subunits might be covalently linked by lysosomes.

Muir: One is reminded of the extra-cellular cross-linking which occurs in collagen, but these linkages are not peptide bonds and perhaps non-peptide links may be formed in lysosomes.

McIlwain: I wonder whether amino acids released by lysosomal protein degradation could be utilised immediately for protein synthesis by adjacent ribosomes?

Dingle: Recirculation of amino acids—that is re-utilisation of amino acids derived from protein degradation for the synthesis of fresh protein has been demonstrated.

McIlwain: Our own studies suggest that in neurones amino acids derived from degraded protein tend to be released at some distance from ribosomes, for example into the extra-cellular space.

The Biochemistry of the Mucopolysaccharidoses

HELEN MUIR

Kennedy Institute of Rheumatology, Bute Gardens, Hammersmith, London, W6 7DW

The mucopolysaccharidoses are a group of recessive diseases characterised by skeletal abnormalities and progressive mental retardation, accompanied by a very high output in the urine of certain sulphated mucopolysaccharides or glycosaminoglycans (GAG). These compounds are also stored in the tissues where abnormal intensely metachromatic cells may be seen when the tissues are properly fixed (Lagunoff, Ross and Benditt, 1962). Lymphocytes (Mittwoch, 1961; Muir, Mittwoch and Bitter, 1962), bone marrow cells (Reilly, 1941; Alder, 1950) and cultured fibroblasts from patients' skin (Danes and Bearn, 1966) are also metachromatic. The metachromatic material in the organs of patients with Hurler's syndrome was identified by Brante (1952) as a sulphated compound resembling chondroitin sulphate. Later the stored material was shown to be heparan sulphate (Brown, 1957; Knecht *et al.*, 1967) or both dermatan sulphate and heparan sulphate (Meyer *et al.*, 1959) and that such patients excreted excessive amounts of these compounds in their urine (Dorfman and Lorincz, 1957).

Many reports have since appeared of the excessive excretion of sulphated GAG in recessive diseases which have a number of common clinical signs. The various forms of the mucopolysaccharidoses have been tentatively classified (McKusick, 1963) according to clinical picture, the type of GAG in the urine and the mode of inheritance (Table 7.1). The abnormal GAG appears in the urine in the same relative proportions as are found in the liver and spleen (Dean *et al.*, 1971) suggesting that the abnormal material in the urine derives from that in the organs.

Current knowledge of the mucopolysaccharidoses has been lately reviewed in a symposium (Bearn, 1969) where possible explanations of the biochemical defect are discussed (Muir, 1969; Dorfman and Matalon, 1969).

TABLE 7.1

Syndrome	*Type*	*Inheritance*	*Urinary GAG*
Hurler	I	Autosomal recessive	Dermatan sulphate Heparan sulphate
Hunter	II	X-linked recessive	Dermatan sulphate Heparan sulphate
Sanfilippo	III	Autosomal recessive	Heparan sulphate
Morquio	IV	Autosomal recessive	Keratan sulphate
Scheie	V	Autosomal recessive	Dermatan sulphate
Maroteaux-Lamy	VI	Autosomal recessive	Dermatan sulphate

The possibility that the mucopolysaccharidoses are lysosomal diseases has been proposed, because they show striking morphological similarities with and some clinical resemblance to other storage diseases where there is a known deficiency of a lysosomal hydrolase. Electron micrographs show single membrane bound vesicles like distended lysosomes in the livers of patients with Hurler's syndrome which resemble vesicles of other storage diseases such as Tay-Sachs disease (Van Hoof and Hers, 1964; Callahan and Lorincz, 1966; Lagunoff and Gritzka, 1966; Haust, 1968). Electron micrographs of cultured Hurler fibroblasts show membrane bound vesicles exactly like those in liver cells (Bartman and Blanc, 1970), while brain biopsies show membranous material in the cytoplasm like that in Tay-Sachs disease (Gonatas and Gonatas, 1965; Aleu *et al.*, 1965). There is a considerable increase in lipid material in the brain particularly in the glycolipids G_{M_1}, G_{M_2} and G_{M_3} (Ledeen *et al.*, 1965; Wallace *et al.*, 1966; Ledeen, 1966; Taketomi and Yamakawa, 1967) one or other of which is stored in different gangliosidoses (Schneck *et al.*, 1969). In the visceral organs the stored material consists of gangliosides in addition to GAG, (Brante, 1952; Borri *et al.*, 1966) so that the metabolism of both GAG and gangliosides is disturbed in the mucopolysaccharidoses which must therefore be interrelated in some way. On the other hand the gangliosidoses do not show excessive storage or excretion of GAG, although cultured fibroblasts from the skin of patients with these diseases may contain increased amounts of GAG (discussed below).

In the past few years a number of deficiencies of lysosomal enzymes have been identified which suggest that the stored substance accumulates because of the lack or deficiency of an appropriate degradative enzyme. Owing to their recessive nature, storage diseases are most likely to arise from a deletion leading to the loss of activity of an enzyme. For example in Tay-Sachs disease there is an accumulation of G_{M_2} ganglioside (see Myant this volume) which has a non-reducing terminal residue of N-acetylgalactosamine. Okada and O'Brien (1969) have found that in brain as well as other tissues there was a complete lack of one of two N-acetylhexosaminidase isoenzymes (Robinson and Stirling, 1965) which could not be explained by the presence of an inhibitor. No other lysosomal glycosidases were deficient, but carriers of the disease showed intermediate levels of activity. Similar results were obtained by Sandhoff (1969) and have been confirmed in further cases (Friedland *et al.*, 1970), while three forms of the disease seem to be distinguishable according to the relative proportions of isoenzymes from brain that are separable on DEAE-cellulose chromatography (Young *et al.*, 1970). If the underlying defect in Tay-Sachs disease is the deficiency of an isoenzyme, it should be normal in other lipid storage diseases, as was in fact the case in generalised gangliosidosis, late infantile amaurotic idiocy and Hurler's syndrome (Okada and O'Brien, 1969).

In Fabry's disease there also appears to be a deficiency of a specific lysosomal hydrolase which removes the terminal galactose of a trihexosyl ceramide (Brady *et al.*, 1967; Kint, 1970), which accumulates in this disease (Sweeley and Klionsky, 1963). Another example is late infantile metachromatic leucodystrophy where cerebroside sulphates accumulate in certain tissues (Austin, 1960; Jatzkewitz, 1960) and there is a deficiency of lysosomal arysulphatase A (Austin *et al.*, 1965; Austin, McAfee and Shearer, 1965). Although ^{35}S-labelled cerebroside sulphate could be taken up from the medium by normal fibroblasts and those from patients, the latter could not degrade it (Porter *et al.*, 1970).

Recent work (Suzuki and Suzuki, 1970) indicates that Krabbe's disease may be another example of the deficiency of a highly specific galactosidase, whereas less specific galactosidase acting on synthetic substrates was not diminished. The deficiency of the specific enzyme was unique to this disease and was not found in other lipid storage diseases such as G_{M_1} and G_{M_2}, gangliosidosis, Gaucher's disease or Hurler's disease.

Such examples encourage the search for a deficiency of a lysosomal hydrolase in the mucopolysaccharidoses and several reports have appeared of partial deficiencies of β-galactosidase in the tissues of patients having the Hurler, Hunter and Sanfilippo syndromes (Types I-III, Table 7.1) (Van Hoof and Hers, 1968; Van Gemund *et al.*, 1968; Öckerman, 1968a, b; Gerich, 1969; McBrinn *et al.*, 1969; Ho and O'Brien, 1969; Hultberg and Öckerman, 1969). However if an enzyme deficiency is to be the primary underlying defect of a given recessive storage disease, the enzyme should be highly specific for the stored compound and the deficiency should be confined to the storage disease in question. These restrictions do not appear to apply to the deficiency of β-galactosidase in the mucopolysaccharidoses.

Several components of β-galactosidase may be distinguished in normal liver by starch gel electrophoresis (Ho and O'Brien, 1969) or gel chromatography (Hultberg and Öckerman, 1969) which differ in pH optima and heat lability (Ho and O'Brien, 1969; Öckerman *et al.*, 1969). Two slow moving components on starch gel electrophoresis were deficient in livers from three types of mucopolysaccharidosis (Ho and O'Brien, 1969) and there was a deficiency of one of the gel-filtration components (Enzyme A) (Hultberg and Öckerman, 1969). In normal liver only this component was active against natural as well as synthetic substrates (Hultberg *et al.*, 1970) which included ceramide-β-galactoside, ceramide-β-lactoside, a glycopeptide prepared from transferrin and keratan sulphate from nucleus pulposus. 'Enzyme A' was shown to be present in the lysosomal fraction and liberated non-reducing terminal galactose from the ceramide-β-galactoside and from keratan sulphate, presumably from the non-reducing end, the quantity liberated being inversely proportional to the chain length of the keratan sulphate (Öckerman, 1970).

Although these results are highly suggestive that a deficiency of an isoenzyme of β-galactosidase may be the defect underlying the mucopolysaccharidoses, there are many difficulties in accepting this explanation. Unfortunately the isoenzyme that is deficient is an exoglycosidase whose properties, as shown so far, do not explain why GAGs accumulate in the mucopolysaccharidoses. In these GAGs the galactose residues are not available to the enzyme because they are not in terminal positions except for a limited number in keratan sulphate. It is therefore not surprising that the β-galactosidase which liberated galactose from the various natural substrates listed above did not act on dermatan sulphate, (Hultberg *et al.*, 1970).

If deficiency of β-galactosidase isoenzyme were the primary defect in the mucopolysaccharidoses, since keratan sulphate is a limited substrate, one would expect it to accumulate rather than the polyuronides dermatan sulphate and heparan sulphate which do not have terminal non-reducing galactose residues. These are most commonly involved in the mucopolysaccharidoses (Table 7.1) whereas the metabolism of keratan sulphate is affected so far as is known only in Morquio's disease, which is a different entity both on clinical grounds and in the pattern of abnormal GAG in the urine.

Another difficulty is that unlike the examples of storage diseases where a deficiency of a highly specific enzyme was confined to a particular storage disease, deficiencies of β-galactosidase of a much greater degree have been found in other diseases where GAGs do not accumulate such as G_{M_1} gangliosidosis (Okada and O'Brien, 1968; Dacremont and Kint, 1968; Van Hoof and Hers, 1968), where it has been shown that the ganglioside labelled with ^{14}C-galactose in the non-reducing terminal position is a substrate for the enzyme in normal tissue (Okada and O'Brien, 1968). This enzyme was almost absent from the liver of a patient with generalised gangliosidosis (McBrinn *et al.*, 1969), whereas only a partial deficiency of this enzyme has been found in patients with mucopolysaccharidoses Types I-III. Although it does not explain the accumulation of GAGs, partial deficiency of β-galactosidase might account for the concurrent accumulation of gangliosides in brains of patients with mucopolysaccharidoses.

Assuming that the catabolism of GAGs is brought about by the action of lysosomal hydrolases, it is remarkable that in many storage diseases including the mucopolysaccharidoses, other lysosomal hydrolases are often several times above normal levels (Van Hoof and Hers, 1968; Öckerman, 1968a, b; McBrinn *et al.*, 1969; Öckerman *et al.*, 1969). This includes hydrolases which could conceivably degrade GAGs, such as hexosaminidase and glucuronidase which might remove alternately N-acetylhexosamine and glucuronic acid from the non-reducing ends of GAG chains. Indeed oligosaccharides recovered from Hurler urines appear to lack non-reducing glucuronic acid terminal groups (Dorfman and Matalon, 1969). Similarly the activity of cathepsins A-D was also several times higher than normal (McBrinn *et al.*, 1969), and cathepsin D is known to degrade the protein moiety of GAG-proteins (see Dingle this volume), the macromolecular form in which GAGs normally occur (Muir, 1969).

A major difficulty in assuming that β-galactosidase is important in the normal breakdown of connective tissue sulphated polyuronides, is the position of the galactose residues which are located in the linkage region towards the reducing terminal of the carbohydrate chain, attaching it through xylose to serine of the core protein. In GAG-proteins a large number of serine residues are glycosylated and carry a polysaccharide chain, so that GAG-proteins are large multichain compounds. The structure of the linkage region containing galactose (Fig. 1) has been conclusively established by the work of Rodén and his co-workers for chondroitin sulphates 4 and 6 and for heparin (Gregory *et al.*, 1964; Lindahl *et al.*, 1965; Rodén and Armand, 1966, Lindahl and Rodén 1966a, b; Rodén and Smith, 1966; Lindahl, 1966a, b, c; Helting and Rodén, 1968). That it is also present in dermatan sulphate is shown by the isolation of the same oligosaccharides and glycopeptides after enzymic and chemical degradation (Fransson, 1968a; Stern, 1968), as were obtained from chondroitin sulphate and heparin. The β-galactosidase that is deficient in patients with mucopolysaccharidoses behaves as an exoglycosidase with natural substrates. The suggestion that it may also have low endo-glycosidase activity (McBrinn *et al.,* 1969; Hultberg *et al.,* 1970) seems unlikely in view of the structure of the active centre of an endoglycosidase such as

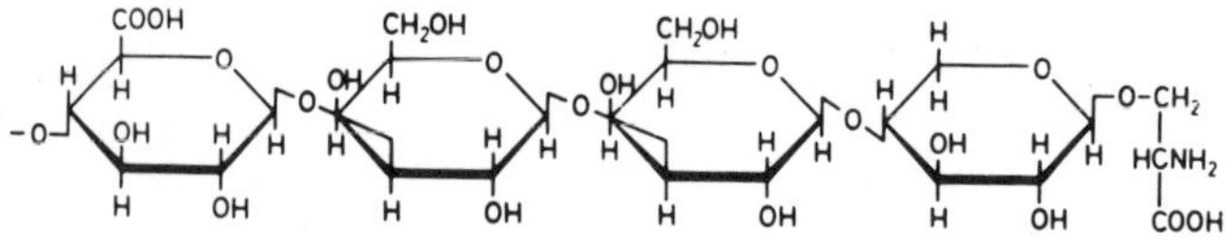

Fig. 1. Structure of the linkage region as established by Rodén and co-workers.

lysozyme which must accommodate four sugar residues to bring about cleavage within the chain and which is therefore inhibited by the corresponding mono, di and trisaccharides (North, 1968). Added to this is the difficulty of explaining why certain GAGs but not all those containing galactose are affected in the mucopolysaccharidoses.

The assumption that there is a defect in degradation needs to be examined. The GAG in pathological organs do indeed appear to be partially degraded. The GAG-proteins of extra-cellular connective tissue are multichain compounds, whereas pathological GAG appear to be free and not attached to protein. Consequently the structural GAG-proteins are extremely difficult to extract intact from tissues, whereas the pathological GAGs are readily soluble in water, so that aqueous fixatives cannot be used. In addition pathological dermatan sulphate (Dorfman, 1964) and heparan sulphate from tissues and urine (Knecht *et al.,* 1967) are patially degraded and deficient in linkage region when compared with the corresponding normal compounds obtained after proteolysis of structural GAG-proteins. On the other hand the GAGs of normal urine, appear to be not more but less degraded than pathological urinary GAG, since they were not notably deficient in linkage region, as estimated by the proportion of xylose residues per chain. (Wasteson and Wessler, 1971). Furthermore, cultured Hurler fibroblasts are capable of degrading exogenous labelled dermatan sulphate as are normal fibroblasts (Dorfman and Matalon, 1969), which would argue against a defect in degradative mechanisms.

Partial deficiency of a β-galactosidase is common to three separate inherited disorders (Types I, II and III), if this deficiency is the primary defect, there is the theoretical difficulty, which McBrinn *et al.* (1969) have pointed out, of explaining how a common enzyme is affected in two autosomal and one X-linked disorder, unless the several isoenzymes that exist (Jungulvala and Robbins, 1968; Furth and Robinson, 1965) are under separate genetic control. Dorfman and Matalon (1969) have also discussed the difficulty of accounting for the existence of two distinct genotypes, one being an X-linked one, which result in similar phenotypes. They suggest that the factor lacking in the mucopolysaccharidoses is a protein having two peptides that are necessary for activity, one peptide being coded for by an autosomal chromosome and the other by an X-chromosome.

Whatever the nature of the factor it affects the metabolism of either dermatan sulphate or heparan sulphate or both in most types of mucopolysaccharidosis (Table 7.1), although there are differences between individual patients in the total amounts and relative proportions of abnormal GAGs, even between brothers having a common mutant gene in Hunter's syndrome (Muir, 1969).

Any comprehensive explanation of the mucopolysaccharidoses is further complicated by the recent finding that the types of GAG involved in these diseases have been extended to include chondroitin sulphate (Muir, 1969; Philippart and Sugarman, 1969). In a patient with the gross clinical picture of Hurler's syndrome, half the GAG in the urine and more than half that stored in the liver was chondroitin sulphate, the remainder being dermatan sulphate (Benson *et al.*, 1971). This situation is quite distinct from the excretion of chondroitin sulphate that is somewhat raised above normal levels in some patients (Berggård and Bearn, 1965; Muir, 1969; Onisawa and Lee, 1970; Murata *et al.*, 1970), which could arise from secondary effects of the presence of dermatan and heparan sulphates in the tissues (Muir, 1969) since they are strong inhibitors of lysosomal hyaluronidase (Aronson and Davidson, 1967) and might therefore interfere with the breakdown of connective tissue chondroitin sulphate that is normally being turned over. This turnover may at the same time be somewhat faster because of the non-specific increase of lysosomal cathepsins in the mucopolysaccharidoses (McBrinn *et al.*, 1969). In connective tissue diseases where there is breakdown of connective tissue the excretion of chondroitin sulphate is greater than normal (Murata *et al.*, 1970).

It appears unlikely that whatever the genetic defects may be, they primarily affect the biosynthesis of GAGs for which prior formation of the core protein is necessary. The formation of dermatan sulphate by either normal or Hurler fibroblasts is inhibited by puromycin (Matalon and Dorfman, 1966) just as is the synthesis of chondroitin sulphate by normal chondrocytes (Telser *et al.*, 1965). This was interpreted as being due to inhibition of core protein formation because direct assay of the enzymes required for the formation of chondroitin sulphate chains showed that their activity was not greatly diminished (Telser *et al.*, 1965). Arrest of or a defect in core protein formation would not therefore lead to an excess of GAG.

Polysaccharide chain synthesis is initiated by the transfer of xylose, the first sugar of the linkage region to the core-protein (Robinson *et al.*, 1966) whose sequence of amino acids around the serine residues probably controls which ones are glycosylated (Baker and Rodén, 1970). Helting and Rodén (1969a, b) have shown how the specificities of the glycosyltransferases that form the linkage region ensure its exact structure (Fig. 1). Using oligosaccharides and glycopeptides of known structure as exogenous competitive acceptors (Tables 7.2 and 7.3)

TABLE 7.2

Substrate Specificity of Galactosyl Transferase for UDP-^{14}C-galactose

Activity % of xylose		*Inactive*
D-xylose	100	L-xylose. D-arabinose
D-xyl-ser	206	galactose
D-gal-xyl	50	gal-lyx
D-gal-xyl-ser	40	gal-gal

TABLE 7.3

Substrate Specificity of Glucuronosyl Transferase for UDP-^{14}C-glucuronic acid

Activity % of gal gal		*Inactive*
1-3		
gal-gal	100	gal-xyl
1-6		
gal-gal	27	xylose
gal-gal-xyl	97	N-acetylgalactosamine
gal-gal-xyl-ser	167	galactose

they found that there were two separate galactosyl transferases for the addition of each galactose of the linkage region, and a glucuronosyltransferase for the addition of the first glucuronic acid which was different from that which added glucuronic acid to the main part of the chain. The product of each successive glycosylation was the acceptor for the enzyme adding the next sugar. A defect in or deletion of any of these enzymes would arrest further chain formation which would lead to a deficiency and not to an accumulation of GAG in the cell.

Fibroblast cultures from patients with storage diseases have been used with some success for diagnosis and the study of underlying defects. Danes and Bearn (1967a) found that fibroblasts from patients with mucopolysaccharidoses developed metachromatic inclusions, and that a high proportion of those from heterozygous carriers did so as well, but cells from fathers of patients with the X-linked disease (Hunter's syndrome) did not (Danes and Bearn, 1967b). It thus became possible to distinguish the X-linked from the autosomal forms of the disease.

In contrast, attempts to detect carriers by abnormal urinary GAG have invariably failed, which suggests that fibroblasts cultures may not be entirely representative of the situation in the body as a whole. Thus the dermatan sulphate isolated from cultured Hurler fibroblasts, after digestion with papain, was of comparable molecular weight with that of normal tissue dermatan sulphate retaining one residue of serine per chain (Matalon and Dorfman, 1968) whereas the dermatan sulphate from the organs of patients had undergone partial degradation of the chain (Dorfman, 1964) as if by the action of lysosomal hyaluronidase. Dermatan sulphate is largely resistant to hyaluronidase but there are some N-acetylgalactosaminyl-glucuronosyl linkages in dermatan sulphate which are susceptible to hyaluronidase (Fransson and Rodén, 1967a) and since these linkages are not randomly distributed along the chain, fragments of all sizes are not formed but mainly small oligosaccharides and large polysaccharides (Fransson and Rodén, 1967b. Fransson, 1968b, c).

Conclusions drawn from the behaviour of cultured fibroblasts should be applied with some reserve in explaining the mucopolysaccharidoses. Fibroblasts from patients with various storage diseases where there is no accumulation of GAG, produced considerably more GAG than normal cells in culture, although the relative proportions of chondroitin sulphate, dermatan sulphate and hyaluronic acid were normal (Matalon and Dorfman, 1969). Cells from Fabry's disease contained abnormal amounts of lipid and glycolipid (Matalon *et al.*, 1969) as well as GAG, suggesting that when lipids of cellular membranes are abnormal, the secretion of GAG may be impaired, so that the cells become metachromatic. Metachromasia of cultured fibroblasts is therefore not specific to the mucopolysaccharidoses. The deficiency of an α-galactosidase (Kint, 1970) in Fabry's disease which is highly specific for the hydrolysis of the terminal galactose residue of the stored trihexosylceramide (Brady *et al.*, 1967) is unlikely to be the direct cause of the accumulation of GAG by these cells.

Skin fibroblasts from patients with mucopolysaccharidoses were distinctive however in producing dermatan sulphate as the predominant GAG (Matalon and Dorfman, 1969) in considerably increased amounts (Matalon and Dorfman, 1966; Schafer *et al.*, 1968). This was so even with fibroblasts from patients with Sanfilippo's syndrome (Type III)

(Matalon and Dorfman, 1969) where the excretion of heparan sulphate alone is abnormal (Table 7.1). Since fibroblasts cultures from none of the mucopolysaccharidoses of Types I-III contained heparan sulphate although its metabolism is disturbed in all three diseases, Matalon and Dorfman (1969) have suggested that skin fibroblasts cannot synthesise this GAG. The metabolism of dermatan sulphate and heparan sulphate would appear to be closely linked and it may be significant that iduronic acid has so far been found in addition to dermatan sulphate only in heparin (Cifonelli and Dorfman, 1962; Radhakrishnamurthy and Berenson, 1963; Wolfrom *et al.*, 1969) to which heparan sulphate is closely related. (Lindahl, 1970, Lindahl, 1966b, Cifonelli, 1970). A disturbance of heparin metabolism in the mucopolysaccharidoses has not been noted however. Fibroblasts from heterozygotes for the Hurler, Hunter and Sanfilippo syndromes contained similar amounts of GAG as the patients' fibroblasts with a predominance of dermatan sulphate (Matalon and Dorfman, 1969).

Skin fibroblast cultures from patients with Morquio's disease likewise did not exhibit the abnormality of keratan sulphate metabolism seen in the patient, as they produced the normal proportions of each GAG although in increased amount (Matalon and Dorfman, 1969). Matalon and Dorfman have suggested that skin fibroblasts cannot synthesise keratan sulphate either. In Morquio's disease the excretion of keratan sulphate is increased together with a moderately raised output of chondroitin sulphate. Keratan sulphate appears to be exclusive to cartilage, nucleus pulpous and cornea and does not possess the linkage region that is common to the other sulphated GAGs. The fact that fibroblasts do not produce all types of GAG, together with the time required for cultures to be set up, make antenatal diagnosis more reliable if the relative proportions and amounts of all the GAGs can be determined in amniotic fluid (Matalon *et al.*, 1970). In one instance the diagnosis made in this way was confirmed after the patient was born. The total GAG concentration in the fluid does however vary during pregnancy (Dane *et al.*, 1970).

The progressive mental retardation seen in the mucopolysaccharidoses, probably results from the accumulation both of gangliosides (Ledeen *et al.*, 1965; Taketomi and Yamakawa, 1967), and GAGs in the brain (Meyer *et al.*, 1959). Heparan sulphate can be formed by glial tumour cells in culture (Dorfman and Ho, 1970) in contrast with skin fibroblasts and it may be significant that Kaplan (1969) has observed that severe mental retardation is associated with the excretion of heparan sulphate.

The connective tissue abnormalities of the mucopolysaccharidoses resemble those seen in generalised gangliosidosis or pseudo-Hurler's disease and may have similar causes. They could be the result of the accumulation of glycolipids in connective tissue cells and possibly also due to the presence of dermatan sulphate in bone and cartilage (Meyer and Hoffman, 1961) where it does not normally occur.

Whether or not the mucopolysaccharidoses are true storage diseases analagous to Tay-Sachs disease has not yet been resolved. That they may be has gained considerable support from experiments with cultured skin fibroblasts. Fibroblasts from patients with mucopolysaccharidoses contained about five times as much GAG as normal cells (Matalon and Dorfman, 1969). There was a greater net synthesis (Schafer *et al.*, 1966) and accumulation of labelled intracellular GAG in Hurler fibroblasts than in normal cells (Matalon and Dorfman, 1968; Fratantoni *et al.*, 1968a), but there was no difference in the amount (Schafer *et al.*, 1966) or rate at which GAG were secreted into the medium (Matalon and Dorfman, 1968; Fratantoni *et al.*, 1968a).

From the kinetics of the uptake and secretion of labelled sulphate by normal and diseased cells in culture Fratantoni *et al.* (1968a) have concluded that in diseased cells there is a defect in the degradation of a separate intracellular pool of GAG which therefore expands continuously and never reaches a steady state as in normal cells (Fig. 2).

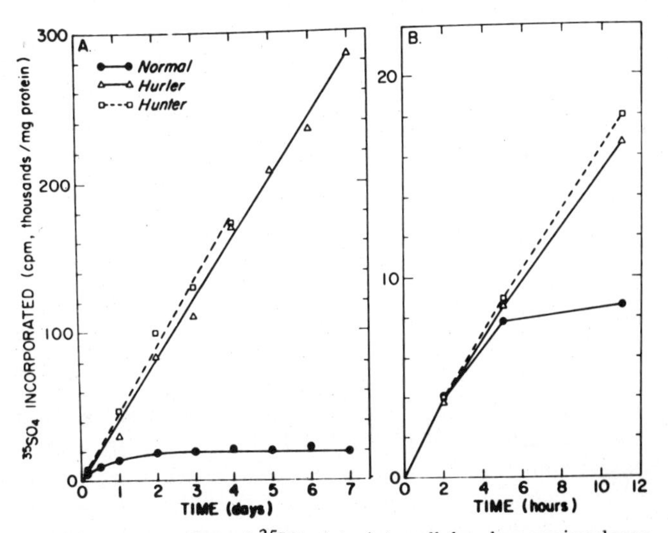

Fig. 2. Incorporation of $^{35}SO_4$ into intracellular glycosaminoglycan (MPS) by fibroblasts from a normal subject and from patients with the Hurler and Hunter syndromes. An expanded plot of the first 12 hours of the experiment is shown in B. The specific activity of the labelled sulphate was 20,000 cpm per μmole. From Fratantoni *et al.* (1968a).

Its rate of formation, however, was normal, as isotope entered the intracellular pool at the same rate as in normal cells. Chase experiments showed that most of the labelled GAG within the cells was released only slowly from Hurler or Hunter cells and appeared in the medium in a dialysable form (Fig. 3). Since the formation of macromolecular GAG in the medium was the same as in normal cells, Fratantoni *et al.*

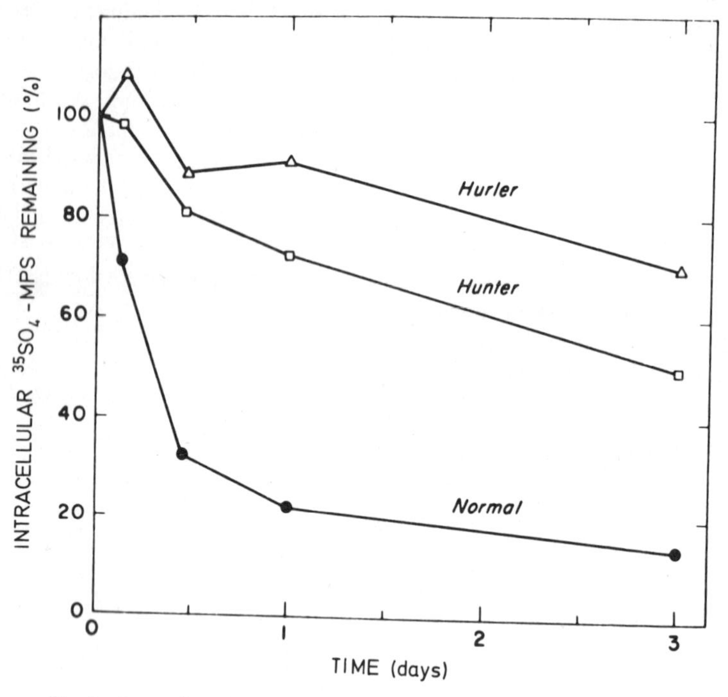

Fig. 3. Loss of radioactivity from intracellular glycosaminoglycan (MPS) during chase with unlabelled medium. From Fratantoni *et al.* (1968a).

(1968a) suggest that there are two intracellular pools, one of small size and rapid throughput where the GAG are destined for export and another containing GAG which are degraded before being released into the medium; only the latter being deranged in Hurler or Hunter cells.

This view has been questioned by Dorfman and Matalon (1969) who point out that these conclusions are based on calculations of the proportion of total radioactivity lost from each type of cell but as the diseased cells are known to contain about five times as much GAG as normal cells, the actual loss of radioactivity was similar in both types of cell. Since dermatan sulphate is the predominant GAG in Hurler or Hunter cells (Matalon and Dorfman, 1969) and since the capacity to degrade dermatan sulphate by liver lysosomes *in vivo* appears to be more limited than the capacity to degrade chondroitin sulphates (Aronson and Davidson, 1968) the degradative enzymes for dermatan sulphate might become easily saturated so that the rate of degradation of the dermatan sulphate would not be proportional to the amount of dermatan sulphate in the cell. If this were so, a slight increase in the rate of formation of dermatan sulphate would lead to accumulation of this GAG. A difficulty in assuming that there is a defect in degradation is that normal and Hurler fibroblasts appear to be equally capable of degrading exogenous labelled dermatan sulphate, prepared from Hurler cells when this was added to the medium (Dorfman and Matalon, 1969). Furthermore the dermatan sulphate obtained by proteolysis of Hurler cells was of similar chain length as normal skin dermatan sulphate retaining one serine residue per chain (Matalon and Dorfman, 1968), although when isolated directly from the cells by disruption it did not appear to be multichain GAG-protein, but single chains without core protein (Fratantoni *et al.*, 1968a) like that in Hurler organs. Perhaps the lysosomal machinery for degrading exogenous and endogenous materials may not be exactly the same.

Whatever the interpretation of their results, Fratantoni *et al.* (1968b); 1969) have made the remarkable observation that cells from a different genotype or culture medium therefrom, correct the storage abnormality of cultured Hurler or Hunter fibroblasts, whereas the cells or medium from different subjects having the same genetic defect do not. (Fig. 4 and 5). Thus Hunter and Hurler cells can mutually correct each other's storage defect even though both diseases involve the same GAGs and are very similar clinically. This finding is compatible with the different recessive inheritance of the two diseases. Similar observations have been extended to include Sanfilippo's syndrome and other types of mucopolysaccharidosis. Classification of the various types can be made on this basis whereby the Scheie syndrome appears to be the same disorder as Hurler's syndrome but Sanfilippo's syndrome is not (Neufeld and Cantz, 1970).

The phenomenon of 'correction' does not require cell contact, but is mediated through macromolecular substances released into the medium or extracted from cells. When added to pre-labelled diseased cells of another genotype the release of labelled sulphate in dialysable form is brought about whereas before treatment the intracellular labelled material was non-dialysable (Fratantoni *et al.*, 1969). When added to diseased cells during incorporation of ^{35}S-sulphate, the accumulation of intracellular label reaches a plateau as it does in normal cells and no

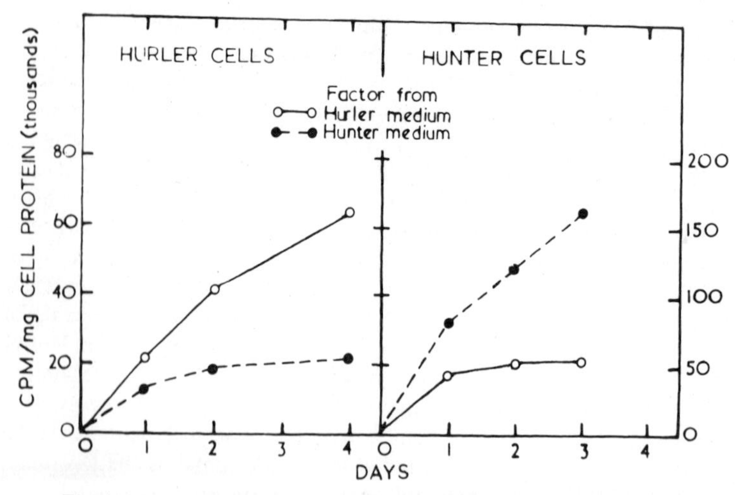

Fig. 4. Accumulation of labelled glycosaminoglycan (MPS) by Hurler and Hunter fibroblasts in the presence of factor preparations. From Fratantoni *et al.* (1969).

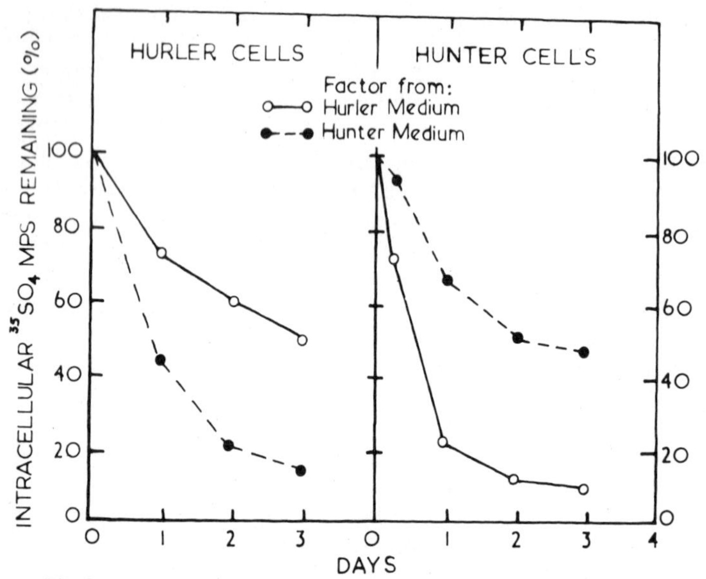

Fig. 5. Chase of labelled glycosaminoglycan (MPS) by Hurler and Hunter fibroblasts in the presence of factor preparations. From Fratantoni *et al.* (1969).

longer goes on increasing linearly over several days (Fig. 4). The half-life of the intracellular labelled material was thus shortened from three days to 22 hours for Hurler cells and from 2 days to 13 hours for Hunter cells, the half-life in normal cells being 8 hours. From these results it would appear that the 'corrective' substances are hydrolases or activators of hydrolases, one of which is lacking in each genetic form of mucopolysaccharidosis, and all of which are needed for the degradation of dermatan sulphate. The factors from Hurler and Hunter cells have been found in particulate as well as supernatant fractions of cell homogenates (Neufeld and Cantz, 1970) but despite the general observation of deficiencies of β-galactosidase in these diseases, the factors were inactive against a synthetic β-galactoside, (Fratantoni *et al.,* 1969). The factor from Hunter cells has been shown to be a macromolecular heat labile substance that is not inactivated by nucleases, and which has negligible absorbance at 260 mμ (Fratantoni *et al.,* 1969). It has been further characterised by polyacrylamide gel electrophoresis, as a protein or bound to a protein of 65,000 molecular weight having a negative charge at pH 8 far lower than nucleic acids or GAGs (Cantz *et al,* 1970).

Since the factor specific for Hunter cells was found in cells from genotypes other than Hunter and that which corrects Hurler cells by genotypes other than Hurler, cultures of normal fibroblasts should contain both factors, as they were found to do thus making the observations internally consistent (Fratantoni *et al.,* 1969). Serum was not needed for the activity of Hunter 'corrective' factor which has now been found in normal human urine together with 'corrective' factors for Hurler and Sanfilippo cells (Neufeld and Cantz, 1970). The factors are thus not merely products of fibroblasts in culture. The Hunter 'corrective' factor in normal cells exhibited hyperbolic dose-response (Neufeld and Cantz, 1970) which may explain why fibroblasts from the heterozygotes accumulate dermatan sulphate in culture (Matalon and Dorfman, 1969) as only half the cells would produce corrective factor, which might not be sufficient under the stress of the unnatural conditions of culture. Material prepared from cells from a Hunter heterozygote contained some corrective factor for Hunter fibroblasts (Fratantoni *et al.,* 1969).

These important results of Neufeld and her co-workers urgently need independent confirmation because they are so far-reaching. Similar studies of other storage diseases might bring to light new aspects of lysosomal physiology, particularly since the macromolecular 'corrective' factors must enter the cell by pynocytosis. Electron micrographs of cells undergoing 'correction' might be informative, as the 'corrective' factors may have some effect on cellular membranes.

In addition to establishing how far actual pool sizes are altered by the 'corrective' factor, several questions need to be examined. It is important to know how long the effect of the 'corrective' factor persists after the factor is removed and whether it affects cells that have not been cultured for long periods, such as lymphocytes which contain metachromatic inclusions *in vivo* (Mittwoch, 1961), that persist in culture (Bowman *et al.*, 1962). Some insight into the process of 'correction' might be found if kinetic studies of the handling of ^{35}S-sulphate by fibroblasts from other storage diseases were examined in the same way when the cultures have begun to accumulate GAG of the normal pattern. Dermatan sulphate is the predominant GAG of fibroblasts from the mucopolysaccharidoses and there may be a more limited capacity to degrade this GAG, as discussed above.

The very high output of non-dialysable GAG in the urine and raised levels in the plasma of patients with mucopolysaccharidoses (Calatroni *et al.*, 1969) are not explicable however if the GAG in the deranged intracellular pool can only emerge when degraded to dialysable fragments, as occurs in fibroblast cultures (Fratantoni *et al.*, 1969). Again the situation in fibroblast cultures is not an exact reflection of the situation *in vivo*.

Nothing is known about what controls the synthesis of GAG-proteins by connective tissue cells, except that the environment, at least of chondrocytes, has a profound influence (Bosman, 1968; Shulman and Meyer, 1968). The artificial conditions of tissue culture could stress the co-ordination of cellular processes which pathological cells might not be able to deal with adequately. This may be the reason why fibroblasts from patients with lipid storage diseases acquire intracellular GAG in culture (Matalon and Dorfman, 1969). Glycolipids are likely to be involved in membrane functions of cells including such synthetic processes as the formation of GAGs, since the glycosyltransferases required for their synthesis are associated with endoplasmic reticulum (Horwitz and Dorfman, 1968). Nevertheless the partial deficiency of β-galactosidase that is so widely observed in the mucopolysaccharidoses remains an enigma.

Dorfman and Matalon (1969) have suggested that β-galactosidase may play some role in the process of secretion by acting on the glycolipids of secretion vacuoles. A partial deficiency of the enzyme might then lead to impaired secretion rather than to a defect in degradation. This, as already pointed out might account for the increase of GAG in fibroblast cultures of lipid storage diseases.

The biochemical defect of the mucopolysaccharidoses remains obscure, despite a great deal of work.

REFERENCES

Alder, A. von (1950). *Schweiz Med. Wchnschr.* **80,** 1095.

Aleu, F.P., Terry, R. O. and Zellweger, H. (1965). *J. Neuropath. and Exper. Neurol.* **24,** 304.

Aronson, N. N. and Davidson, E. A. (1967). *J. biol. Chem.* **242,** 441.

Aronson, N. N. and Davidson, E. A. (1968). *J. biol. Chem.* **243,** 449.

Austin, J. (1960). *Neurology* **10,** 470.

Austin, J., Armstrong, D. and Shearer, L. (1965). *Arch. Neurol.* **13,** 593.

Austin, J., McAfee, D. and Shearer, L. (1965). *Arch. Neurol.* **12,** 447.

Baker, J. R. and Rodén, L. (1970). *Fed. Proc.* **29,** 338.

Bartman, J. and Blanc, W. A. (1970). *Arch. Path.* **89,** 279.

Bearn, A. G. (1969). *Amer. J. Med.* **47,** 661.

Benson, P. F., Dean, M. F. and Muir, H. (1971). In preparation.

Berggård, I. and Bearn, A. G. (1965). *Amer. J. Med.* **39,** 221.

Borri, P. F., Hooghwinkel, G. J. M. and Edgar, G. W. F. (1966). *J. Neurochem.* **13,** 1246.

Bosman, H. B. (1968). *Proc. Roy Soc. Lond. B.* **169,** 399.

Bowman, J. E., Mittwoch, U. and Schniederman, L. J. (1962). *Nature* **195,** 612.

Brady, R.O., Gal, A. E., Bradley, R. M., Märtenson, E., Warshaw, A. L. and Laster, L. (1967). *New Eng. J. Med.* **276,** 1163.

Brante, G. (1952). *Scand. J. Lab. Invest.* **4,** 43.

Brown, D. H. (1957). *Proc. Nat. Acad. Sci.* **43,** 783.

Calatroni, A., Donnelly, P. V. and Di Ferrante, N. (1969). *J. clin. Invest.* **48,** 332.

Callahan, W. P and Lorincz, A. E. (1966). *Amer. J. Path.* **48,** 277.

Cantz, M., Chrambach, A. and Neufeld, E. F. (1970). *Biochem. biophys. Res. Commun.* **39,** 936.

Cifonelli, J. A. (1970). In 'Chemistry and Molecular Biology of the Intercellular Matrix' (ed. E. A. Balazs), Vol. 2, p. 961. London and New York: Academic Press.

Cifonelli, J. A. and Dorfman, A. (1962). *Biochem. biophys. Res. Commun.* **7,** 41.

Dacremont, G. and Kint, J. A. (1968). *Clin. Chim. Acta* **21,** 421.

Danes, B. S. and Bearn, A. (1966). *J. Exper. Med.* **123,** 1.

Danes, B. S. and Bearn, A. G. (1967a). *Lancet* **1,** 241.

Danes, B. S. and Bearn, A. G. (1967b). *J. Exper. Med.* **126,** 509.

Danes, B. S., Queenan, J. T., Gadow, E. and Cederquist, L. C. (1970). *Lancet* **1,** 946.

Dean, M. F., Muir, H. and Ewins, R. J. F. (1971). *Biochem. J.* (in the press).

Dorfman, A. (1964). *Biophys. J.* **4,** 155.

Dorfman, A. and Ho, P. L. (1970). *Proc. Nat. Acad. Sci.* **66,** 495.

Dorfman, A. and Lorincz, A. E. (1957). *Proc. Nat. Acad. Sci.* **43,** 443.

Dorfman, A. and Matalon, R. (1969). *Amer. J. Med.* **47,** 691.

Fransson, L. A. (1968a). *Biochim. biophys. Acta* **156,** 311.

Fransson, L. A. (1968b). *J. biol. Chem.* **243,** 1504.

Fransson, L. A. (1968c). *Arkiv. Kemi* **29,** 95.

Fransson, L. A. and Rodén, L. (1967a). *J. biol. Chem.* **242,** 4161.

Fransson, L. A. and Rodén, L. (1967b). *J. biol. Chem.* **242,** 4170.

Fratantoni, J. C., Hall, C. W. and Neufeld, E. F. (1968a). *Proc. Nat. Acad. Sci.* **60,** 699.

Fratantoni, J. C., Hall, C. W. and Neufeld, E. F. (1968b). *Science* **162,** 570.

Fratantoni, J. C., Hall, C. W. and Neufeld, E. F. (1969). *Proc. Nat. Acad. Sci.* **64,** 360.

Friedland, J., Schneck, L., Saifer, A., Pourfar, M. and Volk, B. W. (1970). *Clin. Chim. Acta* **28**, 397.
Furth, A. J. and Robinson, D. (1965). *Biochem. J.* **97**, 59.
Gemund Van J. J., Giesberts, M. A. H., Gorsira, M. C. B. and Willighagen, R. G. J. (1968). *Maandschr. Kindergeneesk* **36**, 377.
Gerich, J. E. (1969). *New Eng. J. Med.* **280**, 799.
Gonatas, N. K. and Gonatas, J. (1965). *J. Neuropath. and Exper. Neurol.* **24**, 318.
Gregory, J. D., Laurent, T. and Rodén, L. (1964). *J. biol. Chem.* **239**, 3312.
Haust, M. D. (1968). *Exper. Molec. Path.* **8**, 123.
Helting, T. and Rodén, L. (1968). *Biochim. biophys. Acta* **170**, 301.
Helting, T. and Rodén, L. (1969a). *J. biol. Chem.* **244**, 2790.
Helting, T. and Rodén, L. (1969b). *J. biol. Chem.* **244**, 2799.
Ho, M. W. and O'Brien, J. S. (1969). *Science* **165**, 611.
Hoof Van F. and Hers, H. G. (1964). *Compt. Rend.* **259**, 1281.
Hoof Van F. and Hers, J. G. (1968). *Europ. J. Biochem.* **7**, 34.
Horwitz, A. L. and Dorfman, A. (1968). *J. Cell Biol.* **38**, 358.
Hultberg, B. and Öckerman, P. A. (1969). *Scand. J. Clin. Lab. Invest.* **23**, 213.
Hultberg, B., Öckerman, P. A. and Dahlquist, A. (1970). *J. Clin. Invest.* **49**, 216.
Jatzkewitz, H. (1960). *Hoppe-Seylers Z.* **318**, 265.
Jungalvala, F. B. and Robbins, E. (1968). *J. biol. Chem.* **243**, 4258.
Kaplan, D. (1969). *Amer. J. Med.* **47**, 721.
Kint, J. A. (1970). *Science* **167**, 1268.
Knecht, J., Cifonelli, J. A. and Dorfman, A. (1967). *J. biol. Chem.* **242**, 4652.
Lagunoff, D. and Gritzka, T. L. (1966). *Lab. Invest.* **15**, 1578.
Lagunoff, D., Ross, R. and Benditt, E. (1962). *Amer. J. Path.* **41**, 273.
Ledeen, R. (1966). *J. Amer. Vol. Chem. Soc.* **43**, 57 (Review).
Ledeen, R., Salsman, K., Gonatas, J. and Taghavy, A. (1965). *J. Neuropath. and Exper. Neurol.* **24**, 341.
Lindahl, U. (1966a). *Biochim. biophys. Acta* **130**, 360.
Lindahl, U. (1966b). *Biochim. biophys. Acta* **130**, 368.
Lindahl. U. (1966c). *Ark. Kemi.* **26**, 101.
Lindahl, U. (1970). In 'Chemistry and Molecular Biology of the Intracellular Matrix' (ed. E. A. Balazs), Vol. 2, p. 943. London and New York: Academic Press.
Lindahl, U., Cifonelli, J. A., Lindahl, B. and Rodén, L. (1965). *J. biol. Chem.* **240**, 2817.
Lindahl, U. and Rodén, L (1966a). *J. biol. Chem.* **241**, 2113.
Lindahl, U. and Rodén, L. (1966b). *J. biol. Chem.* **240**, 2821.
MacBrinn, M., Okada, S., Woollacott, M., Patel, V., Ho, M. W., Tappel, A. L. and O'Brien, S. J. (1969). *New Eng. J. Med.* **281**, 338.
Matalon, R. and Dorfman, A. (1966). *Proc. Nat. Acad. Sci.* **56**, 1310.
Matalon, R. and Dorfman, A. (1968). *Proc. Nat. Acad. Sci.* **60**, 179.
Matalon, R. and Dorfman, A. (1969). *Lancet* **II**, 838.
Matalon, R., Dorfman, A., Dawson, G. and Sweeley, C. C. (1969). *Science* **164**, 1522.
Matalon, R., Dorfman, A., Nadler, H. and Jacobson, C. B. (1970). *Lancet* **I**, 83.
McKusick, V. A. (1963). 'Heritable Disease of Connective Tissue'. St. Louis: C. V. Mosby Co.
Meyer, K. and Hoffman, P. (1961). *Arthritis and Rheumantism* **4**, 552.
Meyer, K., Hoffman, P., Linker, A., Grumbach, M. M. and Sampson, P. (1959). *Proc. Soc. Exper. Biol. and Med.* **102**, 587.
Mittwoch, U. (1961). *Nature* **191**, 1315.

Muir, H. (1969). *Amer. J. Med.* **47,** 673.
Muir, H., Mittwoch, U. and Bitter, T. (1962). *Arch. Dis. Childh.* **38,** 358.
Murata, K., Ishikawa, T. and Oshima, Y. (1970). *Clin. Chim. Acta* **28,** 213.
Neufeld, E. F. and Cantz, M. J. (1970). *Trans. New Acad. Sci.* Conference on Drug Metabolism. In press.
North, A. C. T. (1968). In 'Fibrous Proteins' (ed. W. G. Crewther), p. 13. Australia: Butterworths.
Öckerman, P. A. (1968a). *Clin. Chim. Acta* **20,** 1.
Öckerman, P. A. (1968b). *J. Lab. Clin. Invest.* **22,** 142.
Öckerman, P. A., Hultberg, B. and Eriksson, O. (1969). *Clin. Chim. Acta* **25,** 97.
Öckerman, P. A. (1970). *Carbohydrate Res.* **12,** 429.
Okada, S. and O'Brien, J. S. (1968). *Science* **160,** 1002.
Okada, S. and O'Brien, J. S. (1969). *Science* **165,** 698.
Onisawa, J. and Lee, T. Y. (1970). *Biochim. biophys. Acta* **208,** 144.
Philippart, M. and Sugarman, G. I. (1969). *Lancet* **2,** 854.
Porter, M. T., Fluharty, A. L., Harris, S. F. and Kihara, H. (1970). *Arch. Biochem. Biophys.* **138,** 646.
Radhakrishnamurthy, B. and Berenson, G. (1963). *Arch. Biochem.* **101,** 360.
Reilly, W. A. (1941). *Am. J. Dis. Child.* **62,** 489.
Robinson, D. and Stirling, J. (1965). *Biochem. J.* **107,** 321.
Robinson, H. C., Telser, A. and Dorfman, A. (1966). *Proc. Nat. Acad. Sci.* **56,** 1859.
Rodén, L. and Armand, G. (1966). *J. biol. Chem.* **241,** 65.
Rodén, L. and Smith, R. (1966). *J. biol. Chem.* **241,** 5949.
Sandhoff, K. (1969). *FEBS Letts.* **4,** 351.
Schafer, I. A., Sullivan, J. C., Svejcar, J., Kofoed, J. Van B. and Robertson, W. (1968). *J. Clin. Invest.* **47,** 321.
Schneck, L., Volk, B. and Saifer, A. (1969). *Amer. J. Path.* **46,** 245 (Review).
Shulman, H. J. and Meyer, K. (1968). *J. Exp. Med.* **128,** 1353.
Stern, E. I. (1968). *Fed. Proc.* **27,** 596.
Suzuki, K. and Suzuki, Y. (1970). *Proc. Nat. Acad. Sci.* **66,** 302.
Sweeley, C. C. and Klionsky, B. (1963). *J. biol. Chem.* **238,** 3148.
Taketomi, T. and Yamakawa, T. (1967). *Japan J. Exper. Med.* **37,** 11.
Telser, A., Robinson, H. C. and Dorfman, A. (1965). *Proc. Nat. Acad. Sci,* **54,** 912.
Wallace, B. J., Kaplan, D., Adachi, M., Schneck, L. and Volk, B. B. (1966). *Arch. Path.* **82,** 462.
Wasteson, A. and Wessler, E. (1971). *Biochim. biophys. acta* (in the press).
Wolfrom, M. L., Honda, S. and Warig, P. Y. (1969). *Carbohydrate Res.* **10,** 259.
Young, B., Ellis, R. B. and Patrick, A. D. (1970). *FEBS Letts.* **9,** 1.

DISCUSSION

Yielding: I was impressed with the similarity which incorporation studies of Neufeld had to situations where there is a defect in control mechanisms. In culture, normal cells grow rapidly at first and incorporate precursors into macromolecules until there is inhibition to

control mechanism which perhaps may be activated by the the medium by cell products. If the control mechanism ak down, cells would keep on growing just as the ier cells appear to do. Moreover, the existence of two mechanisms which had to function at the same time might the reciprocal correction Hunter and Hurler cells.
gle: I should like to point out that failure of digestion of acromolecules within the digestive vesicles does not necessarily mean absence of the appropriate degrading enzymes. It could mean for example that the micro-environment is not suitable for enzyme function; for example Morrison has shown that desulphation of proteoglycans can be inhibited by the presence of sucrose.
Scriver: Has intra-lysosomal pH been measured?
Dingle: This cannot be done by introduction of micro-electrodes, so that one has to depend on indicator dyes, which are not very accurate. However the data available suggest that the intra-lysosomal environment is distinctly acid with a pH of perhaps 3.0 to 5.5. This covers the range of pH optima for the lysosomal enzymes.
Polani: It is interesting that although Hunter and Hurler cell cultures are alleged to complement each other, as evidenced by failure of metachromasia to develop, after complementation, metachromasia occurs in about half the cultured cells from the heterozygous carriers of Hunter's disease. Since this is the X-linked type of Hunter-Hurler syndrome it has been suggested that in heterozygotes metachromasia occurs in cells in which the X-chromosome which is active carries the mutant gene and *vice versa.* It is difficult to understand why heterozygous cells in which the active X-chromosome carries the normal gene don't correct the abnormality in the others.
Robinson: Some recent work of Brew *et al.* (1968, *Proc. Nat. Acad. Sci. U.S.* **59**, 491) demonstrates how protein structure can be involved in specificity of enzyme activity. He showed that UDP-transferase which is normally involved in lactose synthesis consists of two subunits, one of which, the A protein, contains the recognition site for transfer of galactose and the other, which is a directive subunit, identifies the glucose acceptor. This latter subunit appears to be identical with α-lactalbumin. When it is absent, galactose can be transferred by the first subunit to N-acetyl hexosamine as occurs in connective tissue synthesis. I know of no other system where such a mechanism for control of enzyme specificity has been demonstrated.
Yielding: There are several known examples of enzymes which have catalytic subunits and regulatory subunits. Aspartate transcarbamylase is an example. (Gerhart and Schactman, 1965, *Biochem.* **4**, 1054.)

Aspartylglycosaminuria—A New Inborn Error Possibly Due to a Lysosomal Enzyme Defect

R. J. POLLITT and F. A. JENNER

M.R.C. Unit for Metabolic Studies in Psychiatry, University Department of Psychiatry, Middlewood Hospital, P.O. Box 134, Sheffield S6 1TP

The study of aspartylglycosaminuria began four years ago when a thirty-year-old severely subnormal woman was found to be excreting considerable quantities of an unknown ninhydrin-positive compound in her urine. She had been referred to us because she had developed regular periods of mania at about the same time each year, the duration of the manic phase increasing in duration year by year. She had a subnormal brother who was also found to excrete the unknown compound in his urine. He suffered from short episodes of loss of consciousness but did not show any signs of mania (Pollitt *et al.*, 1968). There was a normal brother and the mother and father both appeared normal. The normal members of the family did not have the compound in their urine.

Chemical Studies

The unknown urinary material moved slowly on high voltage electrophoresis at pH 2, and emerged from a column of Sephadex G 15 earlier than the normal urinary amino acids. With ninhydrin-collidine-acetic acid, it stained a characteristic orange-yellow colour very similar to that given by asparagine. By a combination of Sephadex and ion-exchange chromatography, we were able to isolate the material in quantity from the urine. On acid hydrolysis it gave aspartic acid, ammonia and glucosamine in equimolar proportions. This suggested that the compound was 2-acetamido-1-(β^1-L-aspartamido)-1, 2-dideoxy-β-D-glucose (AADG) (Fig. 1a). This was confirmed by the infra-red spectrum, chromatographic properties and eventually by elemental analysis (Jenner and Pollitt, 1967). The systematic name for

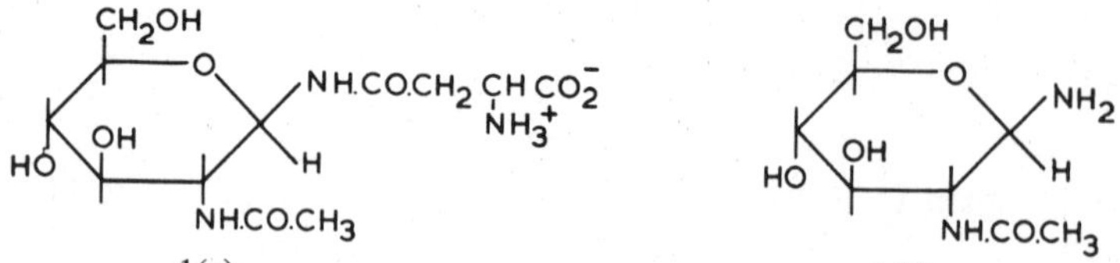

1(a) 1(b)

Fig. 1.

this condition is thus 2-acetamido-l-(β^1-L-aspartamido)-1, 2-dideoxyglycosuria but we found we had to modify this to the easier, though less correct name, of aspartylglycosaminuria.

AADG had not hitherto been found free in nature; it had however been prepared by several workers by degradation of various glycoproteins in which AADG residues form the linkages between the protein chain and the carbohydrate side groups. Our patients were excreting very roughly 300 mg per day of the compound. Traces were detected in the plasma and larger quantities in the cerebrospinal fluid. The excretion persisted in both phases of the female patient's mania and depression and also when she was maintained for five days on a semi-synthetic diet containing no glycoprotein. Much of the excreted AADG was thus of endogenous origin.

The quantitative analysis of AADG presents some difficulties. When AADG was run on our amino acid analyser (Locarte Scientific Company) using the standard procedure of pH 3.2 buffer and an initial temperature of 50°, virtually no ninhydrin positive material was recovered in the effluent. AADG was only recovered from the column when the temperature was lowered and there were considerable losses even at 35°. On the other hand, Micheel and co-workers (1964) have reported qualitative separations of AADG on ion-exchange resins operated at 49.5°, using a pyridine-formate-acetate buffer pH 3.05. In our system AADG is eluted very early (before urea) and has an $\epsilon_{440}/\epsilon_{570}$ ratio of about 0.5. This lability of AADG on ion-exchange resins is also a disadvantage in large scale purifications from urine, though we have now evolved a reasonably satisfactory method involving an initial separation on a large column of Sephadex G 15 followed by a preparative ion-exchange chromatography on a weakly cross-linked sulphonic acid resin, using volatile buffers. AADG from the patient's urine was used as the substrate in the enzyme determinations.

Enzyme Studies

The production of excessive quantities of AADG in our patients could be due to a defect either in glycoprotein synthesis or in degradation. However, the indications are that the carbohydrate residues of glycoproteins are added after the peptide chains have been synthesised, so that AADG is produced free only in the degradation process. The degradation of AADG seems to be the responsibility of a specific enzyme–2-acetamido-1-(L-β^1-aspartamido)-1, 2-dideoxyglucose amidohydrolase, which cleaves the molecule to yield initially aspartic acid and 2-acetamido-1-amino-1, 2-dideoxyglucose (Fig. 1b): the latter compound decomposes spontaneously to yield *N*-acetylglucosamine and ammonia at pH values much above 5 (Makino *et al.*, 1966). This enzyme was first reported from ram epididymis (Murakami and Eylar, 1965), ram testes (Roston *et al.*, 1965) and snails (Kaverzneva, 1965). It has since been found in pig and guinea pig serum (Makino *et al.*, 1966) in many tissues of the rat (Ohgushi and Yamashina, 1968) and other sources. We started to look for this enzyme in readily available human specimens, using a rather crude chromatographic technique. The only sufficiently active source we could find was seminal fluid. That from the (male) patient did not destroy AADG at a significant rate compared with normal seminal fluid, so that we had good presumptive evidence that the basic defect in aspartylglycosaminuria is a deficiency of this AADG amidohydrolase (Pollitt *et al.*, 1968). The very variable nature of seminal fluid with the contributions of the various accessory glands and the obvious limitation to mature males led us to improve our techniques so that the enzyme might be studied in plasma. The biggest difficulty was the relatively low activity of the enzyme, which meant long incubations and high protein concentrations in the reaction mixture. The protein then had to be removed for accurate determination of the released *N*-acetylglucosamine by the Morgan-Elson reaction. Ethanol precipitation proved the most suitable method, though this made the whole procedure rather cumbersome and tedious. However, the results (Fig. 2) showed quite clearly that both patients were deficient in the plasma enzyme; the observed activity was within the error of the method (Pollitt and Jenner, 1969).

The parents and normal brother of the patients all had plasma AADG-ase activities in the centre of the normal range; it would appear that one cannot detect heterozygotes in this way. A similar situation has been found for example, for leukocyte aryl sulphatase-A in some cases of metachromatic leucodystrophy (Percy and Brady, 1968); c.f. Raine (1969).

The enzyme in both rat liver and rat kidney is associated with the lysosomal fractions and shows latency (Ohgushi and Yamashina, 1968; Mahadevan and Tappel, 1967). This lysosomal distribution is consistent with the enzyme's function in tissue glycoprotein catabolism. However, certain differences are noticeable between the various enzyme

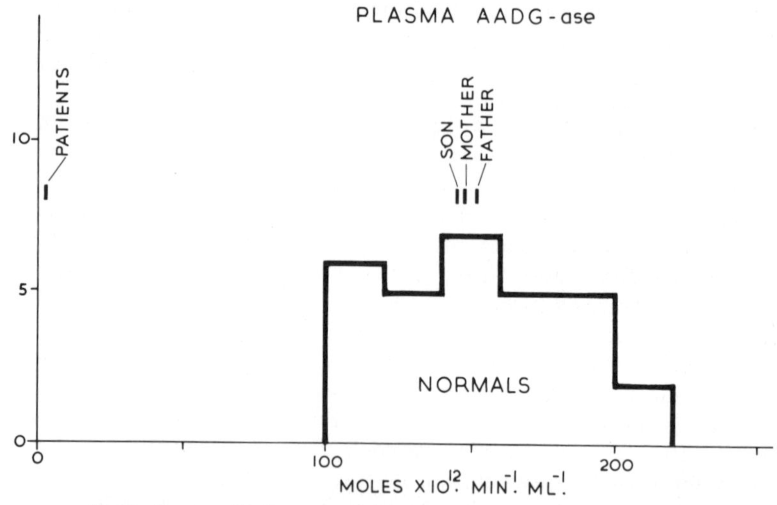

Fig. 2. Enzymatic cleavage of AADG by human plasma. For details see Pollitt and Jenner (1969).

preparations described in the literature. The K_m of the human seminal fluid enzyme working at pH 5.6 is 0.76 mM; the plasma enzyme appears to have a similar K_m, though there was too little activity to determine it accurately. This compares reasonably with the approximate value of 1 mM for the pig serum enzyme (Makino *et al.*, 1966), 0.58 mM for the rat liver enzyme (Conchie and Strachan, 1969) and 0.59 mM for the rat kidney enzyme (Mahadevan and Tappel, 1967). The pH activity of a sample of pooled human seminal fluid resembled fairly closely that of pig serum, though with an increased component in the pH 7-8 region (Fig. 3). This type of curve has also been obtained for a hen oviduct preparation (Tarentino and Maley, 1969). However, the rat liver and rat epididymis preparations prepared by Conchie and Strachan had optimum at pH 7.0 and 8.0 respectively, and the sheep epididymis preparation of Murakami and Eylar also had an optimum of pH 8. The interpretation of these curves is complicated somewhat by the fact that the initial product of hydrolysis is stable at higher pH and gives an enhanced colour with the Morgan-Elson reagent (Makino *et al.*, 1966). These differences obviously warrant further investigation. It may be that each species has more than one enzyme possessing AADG-ase activity in much the same way as there are several β-galactosidases, for example. Nevertheless, it seems reasonable to assume for the time being, that the major enzyme originates in the lysosomes and that this enzyme is missing in our patients.

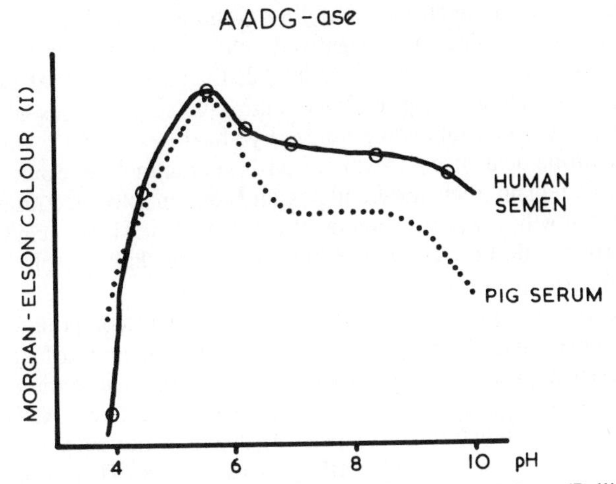

Fig. 3. pH-activity curve for human seminal fluid AADG-ase (Pollitt and Jenner, 1969) and pig serum AADG-ase (Makino *et al.*, 1966).

Possible Storage Aspects

The probable lysosomal distribution of the missing enzyme raises the possibility that there may be some storage aspect to this disease, and Öckerman (1969) has, perhaps prematurely, classed aspartylglycosaminuria as 'glycoprotein storage disease Type II'. There are some interesting findings in the urine which may be relevant to this aspect. In our attempts to purify AADG from the patient's urine it became obvious that there were at least 20 other unusual components present, all of which gave similar colours with ninhydrin but which had higher molecular weights than AADG itself. These presumably had additional carbohydrate residues present and fell into two classes. In the first class the molecular size, judged by the elution position from Sephadex G 15, and the electrophoretic mobility at pH 2 were directly related–the larger molecules having the slower mobility. Compounds in this class presumably consisted of the AADG residue with neutral sugars only added. Electrophoresis showed three major spots in this class probably having one, two and three additional residues, and fainter spots probably containing up to six additional residues. We have made some progress in characterising some of these spots. They seem to contain mainly galactose, mannose and fucose as neutral sugars, but they do not appear to form a homologous series and we cannot be certain yet that each of the electrophoretic bands does not contain a mixture. The second class is more or less neutral at pH 2, but compounds migrate to the anode at pH 4.4; they thus contain acidic groups. The relationship between molecular weight and mobility is more complex in this series; some members having one negatively charged group per molecule, others two, and we can also possibly distinguish positional isomers. Most of these acidic compounds contain sialic acid residues—they are hydrolysed to neutral compounds by heating at 80° with 0.1 *N*-hydrochloric acid or by treatment with neuraminidase. Some of the acidic spots remain unchanged and possibly contain sulphated residues as they move slowly toward the anode at pH 2. This is of interest in view of confirmation (Barker *et al.*, 1969) that the linkage of corneal keratosulphate to protein involves AADG residues.

There are traces of these higher molecular weight compounds in the cerebrospinal fluid. The pattern differs from that found in urine.

There are two possible explanations for the presence of these higher molecular weight compounds. It is possible that the free AADG in the tissues acts as a substrate on which the enzymes normally concerned with glycoprotein side-chain synthesis can doodle. This explanation seems unlikely from what little information we have of the structures of these compounds. More probably they represent some of the side-chains of degraded glycoproteins. As the sialic acid is usually found in the terminal position of such side-chains, it would appear that some of the urinary fragments represent short, complete, undegraded side-chains. In normal humans the aspartic acid residue will have been removed by AADG-ase and the carbohydrate chain would then be further degraded or might be excreted in the urine as such. To what extent the presence of the aspartic acid residue hinders the normal enzymatic degradation of these side-chains is not known: in general carbohydrate chains seem to be degraded from the non-reducing end.

The finding of these higher molecular weight compounds makes the possibility of a storage aspect of this disease seem more likely, though we have no direct proof at this stage. The leukocytes appear normal as might be expected, and other tissues are not as yet available. Clinically there is no clear indication of storage, though one might add that the female patient in particular has rather coarse, sagging facial skin and when we first saw her we did suspect perhaps some type of storage disease. Lysosomal involvement in the central nervous system would provide a convenient explanation of the mental retardation. The concentration of AADG in the plasma and even in the cerebrospinal fluid, is very low and there is as yet no reason to believe that the compound is particularly toxic. AADG-ase activity is present in the rat brain and our findings in the cerebrospinal fluid indicate that it is probably active in the normal human brain. One could imagine compounds, possibly involving lipid-carbohydrate links as well as carbohydrate-aspartic acid links slowly accumulating and interfering with brain function.

A further fifteen cases have recently been reported from Finland (Palo, J., Visakorpi, J. K., Perheentupa, J. and Louhimo, T. (1970) *Scand. J. Lab. Clin. Invest.*, **25**, Supp. 113, 61; Palo, J. and Mattson, K. (1970) *J. Mental Deficiency*, **14**, 168). They show mental retardation, coarse facial features and other signs suggestive of connective tissue disorder. The peripheral lymphocytes are vacuolated.

Summary

Aspartylglycosaminuria is an inborn error of metabolism due to a deficiency of 2-acetamido-1-(β′-L-aspartamido)-1, 2-dideoxyglucose amidohydrolase. This results in the excretion of compounds of a new type (glycoasparagines) in the urine. There may also be storage aspects to the disease. It thus occupies an intermediate position between the classical aminoacidurias and the lysosomal storage diseases. Full elucidation of the link between the enzyme defect and the clinical features must await the discovery of more cases and the opportunity for ultrastructural studies on a variety of organs.

REFERENCES

Barker, J. R., Cifonelli, J. A. and Rodén, L. (1969). *Biochem. J.* **115,** 11P.
Conchie, J. and Strachan, I. (1969). *Biochem. J.* **115,** 709.
Jenner, F. A. and Pollitt, R. J. (1967). *Biochem. J.* **103,** 48P.
Kaverzneva, E. D. (1965). *Izv. Akad. Nauk S.S.S.R., Ser Khim.* **10,** 1911.
Mahadevan, S. and Tappel, A. L. (1967). *J. biol. Chem.* **242,** 4568.
Makino, M., Kojima, T. and Yamashina, I. (1966). *Biochem. biophys. Res. Commun.* **24,** 961.
Micheel, F., Tanaka, Y. and Romer, K. R. (1964). *Tetrahedron Letters* 3913.
Murakami, M. and Eylar, E. H. (1965). *J. biol. Chem.* **240,** PC 556.
Öckerman, P. A. (1969). *Lancet* **1,** 734.
Ohgushi, T. and Yamashina, I. (1968). *Biochim. biophys. Acta* **156,** 417.
Percy, A. K. and Brady, R. O. (1968). *Science* **161,** 594.
Pollitt, R. J. and Jenner, F. A. (1969). *Clin. Chim. Acta* **25,** 413.
Pollitt, R. J., Jenner, F. A. and Merskey, H. (1968). *Lancet* **2,** 253.
Raine, D. N. (1969). *Arch. Dis. Childh.* **44,** 648.
Roston, C. P. J., Caygill, J. C. and Jevons, F. R. (1965). *Biochem. J.* **97,** 43P.
Tarentino, A. L. and Maley, F. (1969). *Arch. Biochem. Biophys.* **130,** 295.

DISCUSSION

Scriver: Has AADG-ase activity been looked for in isolated lysosomes?
Pollitt: Not in humans, though we have found activity in whole leucocytes.
Dingle: Why do you think that there were so many aspartyl oligosaccharides in the urine? I should have thought that they would

have been degraded by enzymes other than AADG-ase. Do you think it possible that there was a deficiency of several degrading enzymes?

Pollitt: We consider that the disorder can probably be explained by a deficiency of AADG-ase alone. Many oligosaccharides are present in normal urine, presumably because they are relatively resistant to degradation. We are able to detect about twenty aspartyl oligosaccharides in patients' urine because we can stain them fairly specifically for the asparagine residue. Possibly the oligosaccharide residues, after removal of the asparagine, are present in urine from normals.

Muir: Do you know whether AADG-ase is active when undegraded glycoprotein is used as the substrate?

Pollitt: The substrate must have both the amino and the α-carboxy group of the asparagine free. Most of the enzyme preparations reported in the literature could act on asparaginyl oligosaccharides as well as on AADG itself.

Myant: I agree with Dr. Dingle that it is strange that there is an accumulation of intermediate oligosaccharides. Is there any evidence that aspartyl oligosaccharides might inhibit activity of other hydrolytic enzymes?

Pollitt: It is possible that the asparagine residue itself may hinder the degradation of the side-chain by glycosidic enzymes.

Dingle: Have you tested whether glycosidic enzymes other than AADG-ase degrade the aspartyl oligosaccharides present in the patient's urine?

Pollitt: No, but we shall probably do this.

Bradford: The mechanism of production of mental subnormality might be due to interference with neurone function or development either because of glycoprotein loss or because of interference with membrane potential since AADG is a charged compound. Such effects might be cumulative over a long period.

Spector: Since AADG is charged, one would expect that this substance does not readily pass from the plasma into the brain. Biochemically induced mental retardation would be expected to be due to an intracerebral abnormality.

Allison: It might be interesting to study the effect of injecting AADG into the cerebral ventricles of a cat.

Polani: Have the corneae been examined by slit lamps? One might find evidence of abnormal storage.

Pollitt: Yes, the corneae appear to be normal.

McIlwain: I was interested to hear that one of your patients had petit mal attacks, because a number of anticonvulsants are ureides and can be regarded as derived from aspartate. Have these patients unusual reactions to anticonvulsants?

Pollitt: No. The male patient is being treated with phenytoin which appears to be effective.

Benson: When you measured AADG-ase activity in the seminal fluid did you find that it was present in the supernatant or only in the spermatozoa, as is the case with the X-type lactic dehydrogenase found only in sperm?

Pollitt: The enzyme activity was measured in the supernatant but the semen had been diluted with an equal volume of water before centrifugation and this would probably have ruptured most kinds of enzyme storage vesicles.

Allison: The enzyme assays were carried out over a considerable period of time. Under these conditions one should be aware that bacterial contamination could give rise to false activity.

Pollitt: All the routine enzyme assays were carried out in the presence of azide in an attempt to counteract this.

Robinson: Has the spectrum of specificity for the sugar next to aspartate been established using synthetic substrates?

Pollitt: No systematic study has been made of this. Only the *N*-acetylglucosamine compound has so far been found naturally.

Allison: Is the *N*-acetylglucosamine the only type of aspartate linkage between carbohydrates and proteins?

Muir: Yes. No other sugar appears to be linked to asparagine. The other type of carbohydrate-protein linkage involves *N*-acetylgalactosamine which is joined to the hydroxyl group of serine or threonine. These linkages are alkali-labile in contrast to the 2-acetamido-1-*N*-β aspartylglucosamine linkages which are relatively stable to alkali (Marshall and Neuberger 1970, *Advanc. Carbohydrate Chem. Biochem.* **25**). Most of the glycoproteins which have small amounts of carbohydrates have the latter type of linkage while submaxillary mucins, blood group substances and connective tissue proteoglycans contain linkages involving hydroxyamino acids.

Raine: Öckerman has shown a raised activity of a number of lysosomal enzymes in tissues and plasma of subjects with mucopolysaccharidosis. This does not appear to happen in the gangliosidoses. Is the reason for this known?

Dingle: When the cell accumulates undigested material, there may be a generalised increase in lysosomal enzyme activity, but this does not always occur. The reason for this is not understood.

Myant: Solubility appears to be an important factor determining whether or not there is storage of substances which accumulate because of an enzyme deficiency. For example, in the lipidoses the undegraded molecules are insoluble and become stored, but when amino acids accumulate, there is usually no storage. I wonder why you expect to find storage for AADG, since presumably this and the urinary oligosaccharides are freely soluble in water?

Scriver: Solubility is probably not the only factor. Cystinosis seems to be a genuine example of a condition where an amino acid is stored in lysosomes. The data of Dr. Schulman and his associates at NIH Bethesda suggest that molecular weight is an important factor in whether a substance can pass through lysosomal membranes. Substances with molecular weight of less than 220-240 seem to be able to diffuse freely but those with higher molecular weights (including cystine) cannot diffuse out and therefore become retained. On a molecular weight basis AADG might well be expected to be too large to pass through the lysosomal membranes and might well be stored.

I should like to ask what is likely to be the nature of the vesicle which contains cystine in cells of patients with cystinosis?

Dingle: Initially they are likely to be storage vesicles, but they could be converted into digestive vacuoles by fusing with primary lysosomes.

Allison: Alan Barrett and John Dingle have shown that a particular lysosomal constituent can bind specifically with certain molecules, notably drugs (Dingle and Barrett, 1969 *Proc. Roy. Soc. B.* **173**, 85). It seems possible that this binding may influence whether or not substances are able to leave the lysosome and I think it is worthwhile considering the possibility that there may be lysosomal constituents to which cystine can be bound.

Dingle: Lysosomal membrane permeability certainly depends on the size of the molecule. For example, molecules larger than sucrose, such as di- and tri-saccharides, can't cross the membrane.

Allison: This is certainly true of polysaccharides. When disaccharides can't be broken down because of the absence of hydrolases large vesicles appear. Work in De Duve's laboratory has shown that proteins in lysosomes are broken down into free amino acids and dipeptides and it seems that these, can cross lysosomal membranes, but not larger peptides (Coffey and DeDuve, 1968, *J. biol. Chem.* **243**, 3255).

Muir: Since oligosaccharides are present in normal urine and since they can't pass through lysosomal membranes, presumably they must have been produced by extralysosomal digestion of polysaccharides.

Dingle: Not necessarily. It has been shown that large molecules, such as dextran, which is a polysaccharide for which there are no cellular degrading enzymes, become liberated into the extracellular space.

Bradford: If there is a restriction on the permeability of lysosomal membranes to large molecules, is it considered unlikely that proteins which travel down the axon to the nerve terminal are able to enter lysosomes? If this is so are proteins degraded in the cytoplasm?

Allison: There is no reason why proteins cannot be engulfed by membrane invagination and enter the lysosomes this way. I know of no evidence that proteolytic enzymes are ever released into the cytoplasm under normal conditions.

Dingle: I think one should distinguish between two types of vesicles. First the true lysosomes in which ingested material is normally degraded by enzymic hydrolysis. This is the type which will accumulate undegraded material when there is deficiency of hydrolytic enzyme. The second type of vesicles are not lysosomes and contain other substances, such as hormones, for export outside the cell.

Scriver: I suspect that lysosomal membranes may differ from plasma membranes in composition. Is there much known about this?

Dingle: Lipid analyses have been carried out and differences demonstrated but one should remember that different types of vacuoles are derived from various sources and one would expect them to have different compositions, e.g. secondary vacuoles are derived from the plasma membrane and thus have fatty acid compositions similar to that of plasma membrane. On the other hand vacuoles which are derived from the Golgi apparatus consist of smooth membrane have fatty acid compositions which differ from that of the plasma membrane.

Robinson: Some lysosomal enzymes such as esterases appear to be tightly bound on to the membrane and are not released when the lysosomes are disrupted.

Allison: One has to be very careful in the interpretation of studies on enzyme localisation using subcellular fractionation since artefacts may easily be introduced. Indigogenic esterase studied by Stanley Holt appears to be particulate in histo-chemical preparations but is released into the supernatant fraction after homogenisation.

Spector: Do you think that non-membrane-bound lysosomal enzymes are structurally organised or just present randomly in solution?

Dingle: I do not know of any evidence about this point but I would think that in digestive vacuoles enzymes in the aqueous medium are not organised, but the situation may be different in primary lysosomes where organisation would seem quite possible.

Polani: When digestive vesicles fuse with primary lysosomes is it thought that the contents of the former are just injected into the latter or do the membrane structures of both vesicles become assembled into one larger vesicle?

Dingle: There is strong evidence from EM pictures that membranes from both types of vesicles fuse and that both contribute to the final structure.

Allison: In the yolk of insect eggs one can demonstrate the formation of vesicles containing yolk material. These vesicles can be stored for quite long periods before enzymes in them suddenly become activated and proceed with digestion of yolk, so that the products of hydrolysis can be used for synthesis by the developing egg (Allison and Hartree, 1969. *Research in Reproduction* **1**, 206).

Robinson: I have noticed that p-nitrophenyl derivatives enter isolated lysosomes much more easily than glycerophosphates.

Scriver: Since certain substrates can enter lysosomes presumably there are recognition sites which allow them access and one would expect recognition sites to exist also on the inside of the lysosome. Since in some cases however substances can't leave the lysosome, perhaps the recognition site becomes altered, possibly by auto digestion.

Allison: Yes, it is quite likely that some change occurs, perhaps a conformational change.

Dingle: When vacuoles are first formed by being nipped off the membrane, they readily fuse with other vacuoles. Later, however, they become smaller and, possibly because of dehydration, their contents become more concentrated and they then may lose the property of being able to fuse with other vacuoles.

Muir: One wonders whether Neufeld's corrective factor might allow the invaginated structure to recognise and fuse with lysosomes. Since the corrective factor is a protein of about 65,000 M.W. it would need to enter the cell by a plasma membrane invagination.

Benson: There is evidence from electron micrographs that metachromatic granules found in cultivated fibroblasts derived from skins of subjects with Hunter-Hurler syndrome are storage material bound by membrane and have been identified as lysosomes (van Hoof and Hers, 1964; *C.R. Acad. Paris* **259**, 1281; Lagunoff and Gritzka, 1966, *Lab. Invest.* **15**, 1578). Has anyone any views on why similar metachromatic granules may be observed under similar conditions in cultured fibroblasts derived from about 7% of apparently normal individuals? One possibility is that these latter may be heterozygous carriers, e.g. for cystic fibrosis.

Muir: In normal guinea pigs certain cells, known as Kurloff cells, contain a chondroitin sulphate-protein in membrane-bound vesicles (Dean and Muir, 1970, *Biochem. J.* **118**, 783).

Allison: Yes, but in culture there is considerably more pinocytosis going on than under physiological conditions and it is not surprising that there is an excess of lysosomal structures.

Polani: We have observed that metachromasia appears more readily in cells which have been previously frozen. One wonders whether freezing could damage lysosomal membranes.

Allison: Possibly, but I should have expected the damage to have been temporary and not to persist during continued culture.

Types of Lysosomal Abnormality and the Light They Throw on Normal Lysosomal Functions

A. C. ALLISON

Division of Cell Pathology, M.R.C. Clinical Research Centre, Northwick Park Hospital, Harrow

Three main classes of congenital abnormality affect lysosomal structure and function.

Lysosomal Enzyme Defect

The first is failure to synthesise a lysosomal enzyme in normal amounts with accumulation of the substrate in lysosomes. The classical example is type II glycogenosis in which glycogen can be mobilised normally although it accumulates within membrane-bounded vacuoles in the tissues. The patients respond to glucagon administration by a rise in blood glucose. The disease is explained by a complete lack of α-glucosidase (Hers, 1963) which breaks down glycogen to glucose and is one of the many hydrolases present in lysosomes. Thus extra-lysosomal breakdown of glycogen can occur but if glycogen enters the lysosomal vacuolar system—by autophagy or some other process—it remains undigested and so accumulates. The relevance of this type of lysosomal defect to the mucopolysaccharidoses and gangliosidoses is discussed by other speakers at this meeting.

Chediak-Higashi Syndrome

The second type of defect involving lysosomes is illustrated by the Chediak-Higashi syndrome in man and comparable conditions in mink, cattle and mice. All are found in individuals homozygous for an autosomal mutant gene. The Chediak-Higashi syndrome was recognised because of the presence of unusually large granules in peripheral blood leucocytes (Beguez-Cesar, 1943; Steinbrink, 1948; Chediak, 1952; Higashi, 1954). The large granules are seen in a proportion of cells belonging to many different types. Thus, some lymphocytes have large azurophilic granules, whereas some of the granules of polymorphs, eosinophils and basophils show the appropriate specific staining reactions but are unusually large. Thus the condition has been accepted as gigantism of the usual granules rather than accumulation of a non-metabolisable substance in lysosomes (Bessis *et al.*, 1961).

One of the features of the disease is dilution of skin, hair and eye colour. Mink homozygous for the Aleutian gene have the commercially valuable pale 'mutant' fur, cattle with the condition are partial albinos, Beige mice have dilute coat colour (Lutzner *et al.*, 1967), and children show corresponding defects. In children of fair parents the pigment dilution can be so marked that it resembles albinism, sometimes associated with photophobia and nystagmus. However, in children with darker complexions (as in the early case reports from Cuba and Japan), there is irregularly grey hair and mottled skin pigmentation. Windhorst *et al.* (1966) found that the melanin granules in the skin of children with the Chediak-Higashi syndrome are also abnormally large, so that they absorb light less efficiently than the same amount of pigment in smaller granules. Thus the basic defect in the syndrome appears to be in the membranes surrounding cytoplasmic organelles, which are either synthesised unusually large or fuse with one another to form giant granules. Several investigations, using ultrastructural histochemistry, have shown that the distinctive feature of the Chediak-Higashi syndrome is the presence of giant, single, membrane-bound, lysosome-like organelles with pleomorphic contents. These have been described in haemopoietic tissue, skin, ocular pigmented epithelium, peripheral nerve, hair, adrenal, pituitary and organs of the gastrointestinal tract (White, 1967).

It is not yet clear how the presence of these large inclusions is related to the increased susceptibility shown by affected children to infections such as pneumonia, oral ulceration or pyoderma. Mink homozygous for the recessive gene (*aa*) also show increased susceptibility to Aleutian disease, which is characterised by positive antiglobulin tests (reactive with a broad spectrum of globulins), anaemia, marked plasma cell proliferation, increased gamma globulin levels and fibrinoid vascular lesions, mainly in the kidneys. Aleutian disease is transmissible by cell-free filtrates and is thought to be due to a virus that passes through a 50 mμ millipore filter and is ether-resistant (Eklund *et al.*, 1968). Some of the affected mink show myeloma-like, monoclonal IgG components in their serum (Porter *et al.*, 1965; Williams, 1968). The renal and other vascular lesions appear to be due to deposition of virus antigen-antibody complexes in the renal glomeruli and other sites (Porter *et al.*, 1969).

Further comparative studies of the Chediak-Higashi syndrome of humans, mink and cattle were reported by Padgett *et al.* (1967). Ingestion by leucocytes and intracellular destruction of bacteria was not perceptibly different from normal, although the large granules sometimes failed to discharge their contents into phagosomes. Accelerated turnover of spingolipids in leucocytes from Chediak-Higashi patients has been described (Kanfer *et al.*, 1968), together with excessive intramedullary granulocyte destruction; however the evidence for the latter is increased serum muramidase (lysozyme) activity and is indirect. No abnormalities have yet been detected in concentrations of lysosomal enzymes or of humoral antibodies in humans or animals with the syndrome (Padgett, 1968). Thus the chemical basis of the striking morphological defect is unknown, as is the way in which it is related to increased susceptibility to infections.

Another peculiarity of the Chediak-Higashi syndrome in man is the tendency of affected children, at some time ranging from a few months of age to adolescence, to develop a progressive or malignant type of disease. Characteristically there is fever, hepatosplenomegaly, lymphadenopathy, anaemia, thrombocytopenia, granulocytopenia and often jaundice. Several patients with splenomegaly have shown mononuclear cell infiltration and other features suggestive of malignant lymphoma (Efrati and Jonas, 1958; Saraiva *et al.*, 1959; Dent *et al.*, 1966). The myeloma-like features of Aleutian disease have already been mentioned. This is associated with a virus infection, but there is not yet any convincing evidence that the same is true of the malignant phase of the Chediak-Higashi syndrome in man. Long-term cultures of lymphocytes established from patients have shown the characteristic large lysosome-like inclusion bodies (Douglas and Fudenberg, 1969). In view of the finding of large secondary lysosomes of nerve cells of Chediak-Higashi patients, the possible involvement of the nervous system must be considered, but children do not show mental deficiency.

Chronic Granulomatous Disease of Childhood

This is a condition in which the anti-microbial action of certain specialised lysosomes, those in neutrophils and monocytes, is defective for reasons which are not yet fully understood. The clinical condition is characterised by eczema, lymphadenopathy, hepatosplenomegaly and recurrent suppurative or granulomatous infections (Bridges *et al.*, 1959; Carson *et al.*, 1965). The disease characteristically affects male children and in its extreme form is usually fatal in the first decade of life. The children are affected by low-grade pyogenic pathogens. Gamma globulins are usually elevated and there is no defect in antibody production. Radiological examination often shows chronic pneumonia with hilar node enlargement and abscess formation. Osteomyelitis is frequent, as is the presence of calcified granulomatous lymph nodes in the abdomen and neck. A consistent pathological feature is the presence of lipid-laden histiocytes in the liver and lymph nodes.

The development of *in vitro* test systems of phagocyte function revealed that neutrophils from patients with chronic granulomatous disease have a normal capacity to ingest bacteria but defective intracellular killing (Quie *et al.*, 1967). Windhorst *et al.* (1968) reported that chronic granulomatous disease is inherited as an X-linked recessive with full manifestation in males and a partial defect in bacterial killing in the cells of heterozygous mothers. Another *in vitro* assay system, developed by Baehner and Nathan (1968), depends on the reduction of the dye nitroblue tetrazolium (NBT) during phagocytosis of latex particles. Patients with chronic granulomatous disease show marked impairment in dye reduction whereas heterozygous mothers usually show reduction intermediate between that of patients and of normal subjects. However, the expression in heterozygotes is highly variable, and can be either severe or slight. Defective killing of bacteria is found in the monocytes of patients as well as their neutrophils (Davis *et al.*, 1968b; Good *et al.*, 1968).

Early electron micrographs of chronic granulomatous disease neutrophils suggested that the morphological events related to phagocytosis were abnormal (Davis *et al.*, 1968b), namely formation of a phagocytic vacuole (vacuolisation) and release of lysosomal enzymes (degranulation) (Quie *et al.*, 1967). However, other studies have not confirmed the presence of any abnormality in this respect (Davis *et al.*, 1968a).

Metabolic studies of neutrophils from chronic granulomatous disease patients have shown a decreased respiratory burst, decreased glucose 6-phosphate oxidation and decreased peroxide formation during phagocytosis (Holmes *et al.*, 1967). Baehner and Karnovsky (1968) have reported a decreased NADH oxidase activity in the neutrophils of the patients. This enzyme is important in the cyanide-insensitive respiratory burst during phagocytosis and may be related to generation of hydrogen peroxide. However it is not yet clear whether this enzyme is consistently reduced in activity, and the biochemical defect leading to decreased intracellular killing is not yet defined.

Several families have been described with defective killing of bacteria by leucocytes which do not conform to the classical chronic granulomatous disease pattern. In a number of these females have shown the clinical and laboratory manifestation of the disease (see Douglas and Fudenberg, 1969), and this has led to the suggestion that the condition can sometimes be inherited as an autosomal recessive. Two families have been described with two red-headed girls in each family affected with recurrent cold suppurative infections due to pyogenic staphylococci; this has been termed Job's syndrome (Davis *et al.*, 1966; Bannantyne *et al.*, 1969). In two members of one family examined neutrophils showed defective killing capacity and tetrazolium reduction. The general interest of these conditions is that they emphasise the importance of leucocytes in protection against bacterial infection even under conditions when antibody formation is normal and patients are treated with antibiotics. In other patients with recurrent infections leucocyte function is normal but plasma opsonising factors are missing (Davis *et al.*, 1968a; Alper *et al.*, 1968). Thus the co-ordinated action of all these factors is required along with leucocyte degranulation for protection against organisms, even those that are only weakly pathogenic under normal conditions.

Lysosomal Inclusion Bodies Due to Substrate Saturation

In the first type of congenital defect mentioned above inclusion bodies develop because normal amounts of substrate enter the lysosomal system and one of the enzymes required for their breakdown is missing. A second type of inclusion body is formed when the lysosomal system is overloaded with substrate. An instructive example is found when a protein has two or more polypeptide chains which are independently synthesised and when the synthesis of the two chains is unbalanced. In normal erythrocyte precursors there is balanced synthesis of α, β and γ-chains of haemoglobins. In α-thalassaemia synthesis of the α-chain is diminished, and there is a relative over-production of γ-chains in the foetus and of β-chains in the adult. These combine with each other to form tetramers known as γ_4 and β_4 haemoglobins which are relatively stable (though less so than the normal haemoglobins) and are found in the peripheral blood. In β-thalassaemia there is no free α-chain component in peripheral blood, from which it was first thought that free β-chains might be required to release newly synthesised α-chains from polyribosomes. However, Fessas (1963) showed that the erythroblasts and erythrocytes in thalassaemia contain prominent inclusion bodies, which were later isolated and shown by Fessas *et al.* (1966) to consist largely of α-chains.

Thus in β-thalassaemia it is clear that excess α-chains are released from polyribosomes but are very unstable and precipitate to form the inclusions observed in erythroblasts, normoblasts and erythrocytes in bone marrow and—particularly after splenectomy—in peripheral blood. We have examined such cells by light and electron microscopy to see whether the inclusions are lysosomal. In erythroblasts and normoblasts of patients with β-thalassaemia the inclusion bodies containing denatured haemoglobin α-chains are surrounded by single membranes and give histochemical reactions for acid phosphatase and other lysosomal enzymes. They also give periodic acid Schiff reactions. Quite often such vacuoles contain pleiomorphic inclusions, and sometimes they appear to be adherent to the cell membrane. In some mature erythrocytes inclusion bodies are not surrounded by membranes but are adherent to the cell membrane. These resemble Heinz bodies formed in mature erythrocytes in the presence of certain drugs. The membrane of such cells is apparently altered (perhaps by association with denatured protein) so that erythrocytes containing inclusions can be recognised by the spleen as abnormal and eliminated from the circulation.

Thus a class of abnormalities exists in which unstable proteins or other products are synthesised, become denatured and are secondarily incorporated into lysosomes. The basic defect is not failure of breakdown of substrate incorporated in normal amounts, as in the examples discussed above, but overloading of lysosomes with amounts of material so large that they cannot be digested by the enzymes available. We have considered the case in which an unstable product arises through failure of synthesis of a complementary polypeptide chain which normally stabilises it. Unstable products might well arise in other ways, for example if a protein were stabilised by attaching to it a carbohydrate or some other moiety.

Lysosomes in Protein Synthesis and Turnover

In maturing erythrocytes of β-thalassaemia it is clear that selective degradation of α-chain takes place. J. Banks (private communication) has found that if the cells are allowed to incorporate radioactive amino acids and are incubated further in cold medium, label is lost from α-chains much more rapidly than from β-chains. The question arises whether the lysosomes might play a similar role in disposing of much smaller excesses of polypeptide chains appearing in the course of normal synthesis. If an efficient clearing mechanism of this sort exists, it would be unnecessary to have exactly balanced synthesis of polypeptide chains of proteins containing more than one type of chain. The β-thalassaemia example is instructive for another reason. The polypeptide chains of haemoglobin are synthesised on polyribosomes lying free in the cytoplasm, not membrane-bound, and yet the denatured α-chains in erythrocyte precursors can find their way into the lysosomal vacuolar system. One of the hallmarks of denaturation is that non-polar groups of polypeptides which are usually buried within the folded molecule become exposed at the surface, so it is not perhaps surprising that denatured proteins should have increased affinity for membranes.

This point is of interest in another problem, the breakdown of cytoplasmic proteins that must accompany the normal turnover which is nicely revealed by radioactive tracers. How can hydrolytic enzymes within the lysosomal vacuolar system gain access to cytoplasmic proteins? By analogy with the argument just presented it is conceivable that the cytoplasmic proteins first become denatured and then attached to the membranes of the vacuolar system, after which by a process of molecular rearrangement they can actually enter the vacuoles and be digested by the lysosomal hydrolases.

To ascertain whether such a mechanism is feasible I have carried out some experiments with bovine serum albumin (BSA). Armour crystalline BSA was trace labelled with ^{125}I and ^{131}I by the chloramine T method (which is thought to produce minimal damage), and submitted to high-speed centrifugation (105,000 g for 1 hour). It still contained some partially denatured molecules rapidly removed by the Kuffer cells of the liver after intravenous injection into rabbits or rats; the remaining undenatured molecules were removed slowly until the onset of immune elimination. Rat serum removed after the initial phase of rapid ^{125}I-labelled BSA elimination (presumably containing few or no denatured molecules) was then compared with normal rat serum to which ^{131}I-labelled BSA (containing some denatured molecules) had been added. Both materials were incubated for one hour at 32° with a rat liver lysosomal light-mitochondrial preparation and then centrifuged. In all experiments a much higher proportion of ^{131}I than of ^{125}I was found associated with the sedimented lysosomes. These results suggest that partially denatured protein, which is not aggregated into a sedimentable macromolecular complex, has a higher affinity for the membranes of large-granule organelles than has undenatured protein. The role of this process in protein degradation and turnover is of course speculative, but the mechanism is plausible and requires further exploration. In nerve cells there is considerable turnover of protein and this must imply the existence of exocytosis or degradation. Possibly the numerous lysosomes in the cell body, and the smaller number of lysosomes at the axonal extremity play a part in degradation, and it is not stretching imagination to suppose that disturbance of this system might be associated with abnormal neuronal functioning.

REFERENCES

Alper, C. A., Abramnsen, N., Johnson, R. B., McCall, C. E., Jandl, J. H. and Rosen, F. S. (1968). *J. clin. Invest.* **47,** 1a.

Baehner, R. L. and Karnovsky, M. L. (1968). *Science* **162,** 1277.

Baehner, R. L. and Nathan, D. (1968). *New Eng. J. Med.* **278,** 971.

Bannantyne, R. M., Showran, P. N. and Weber, J. L. (1969). *J. Pediat.* **75,** 236.

Beguez-Cesar, A. (1943). *Boln. Soc. cub. Pediat.* **15,** 900.

Bessis, M., Bernard, J. and Seligman, M. (1961). *Nouv. revue fr. Hemat.* **1,** 422.

Bridges, R. A., Berendes, H. and Good, R. A. (1959). *A.M.A.J. Dis. Child.* **97,** 387.

Carson, M. J., Chadwick, D. L., Brubaker, C. A., Cleland, R. S. and Lanching, B. H. (1965). *Pediatrics* **35,** 405.

Chediak, M. (1952). *Revue Hemat.* **2,** 362.

Davis, S. D., Schaller, J. and Wedgwood, R. J. (1966). *Lancet* **1,** 1013.

Davis, W. C., Huber, H., Douglas, S. D. and Fudenberg, H. H. (1968a). *J. Immunol.* **101,** 1093.

Davis, W. C., Douglas, S. D. and Fudenberg, H. H. (1968b). *Ann. int. Med.* **69,** 1237.

Dent, P. B., Fish, L. A., White, J. G. and Good, R. A. (1966). *Lab. Invest.* **15,** 1634.

Douglas, S. D. and Fudenberg, H. H. (1969). *Med. Clinics of North America* **53,** 903.

Efrati, P. and Jonas, W. (1958). *Blood* **13,** 1063.

Eklund, C. M., Hadlow, W. J., Kennedy, R. C., Boyle, C. C. and Jackson, T. A. (1968). *J. infect. Dis.* **118,** 510.

Fessas, Ph. (1963). *Blood* **21,** 21.

Fessas, Ph., Loukopoulos, D. and Kaltsoya, A. (1966). *Biochim. biophys. Acta* **134,** 430.

Good, R. A., Quie, P. G., Windhorst, D. B., Page, A. R., Rodey, G. E., White, J., Wolfson, J. and Holmes, B. H. (1968). *Seminars in Hematology* **5,** 215.

Hers, H. G. (1963). *Biochem. J.* **86,** 11.

Higashi, O. (1954). *Tohoku J. exp. Med.* **59,** 315.

Holmes, B., Page, A. R. and Good, R. A. (1967). *J. clin. Invest.* **46,** 1422.

Kanfer, J. N., Blume, R. S., Yankee, R. A. and Wolff, S. M. (1968). *New Eng. J. Med.* **279,** 410.

Lutzner, M. A., Laurie, C. T. and Jordan, H. W. (1967). *J. Heredity* **58,** 299.

Padgett, G. A. (1968). *Adv. in Vet, Science* **12,** 239.

Padgett, G. A., Reiquam, C. W., Henson, J. B. and Gorham, J. R. (1967). *J. Path. Bact.* **95,** 509.

Porter, D. D., Dixon, F. J. and Larsen, A. E. (1965). *Blood* **25,** 736.

Porter, D. D., Larsen, A. E. and Porter, H. G. (1969). *J. exp. Med.* **130,** 575.

Quie, P. G., White, J. G., Holmes, B. and Godd, R. A. (1967). *J. clin. Invest.* **46,** 668.

Saraiva, L. G., Azevedo, M., Correa, J. M., Carvallo, G. and Prospero, J. D. (1959). *Blood* **14,** 1112.

Steinbrinck, W. (1948). *Dt. Arch. klin. Med.* **193,** 577.

White, J. G. (1967). *Blood* **29,** 435.

Williams, R. C., Jr. (1968). *Arth. Rheum.* **11,** 593.

Windhorst, D. B., Zelickson, A. S. and Good, R. A. (1966). *Science* **151,** 81.

Windhorst, D. B., White, J. G., Dent, P. B., Decker, J. and Good, R. A. (1968). In 'Immunologic Deficiency Diseases in Man' (ed. D. Bergsma). National Foundation Publication **4,** 424.

DISCUSSION

Yielding: Wouldn't you expect protein denaturation which encourages membrane interaction to be irreversible?

Allison: I envisage that intracellular organelles move around and collide with each other and rebound because of elastic properties of their membranes. Where the protein in the lysosomal membrane is denatured one can visualise the rebound might not occur, and that membranes involved in the collision might fuse.

Muir: Does your model allow for immunospecific reactions of glycoproteins to occur inside lysosomes?

Allison: Yes, in storage type vesicles where glycoprotein is undegraded but not in digestive lysosomes where glycoprotein would be broken down.

Dingle: One can demonstrate that antibodies are formed against substances ingested by pinocytotic vesicles during the intervals before they are degraded. Another illustration of initial integrity of ingested protein molecules is the demonstration that ingested peroxidase continues to be active inside pinocytotic vesicles.

Polani: In the case of β-thalassaemia is the conformational change already built into the fact that you have four α-chains?

Allison: The conformational change is built into the fact that while normally there is a very intimate interaction between α and β-subunits, in the artificial situation where only α-subunits are present, they seem unable to interact.

Polani: Do the α-chains ever crystallise inside the cells?

Allison: No, they are amorphous. We have never seen crystals.

Jepson: Failure of the α-chains to polymerise with themselves is also a property of their structure.

Yielding: Have lysosomes been studied in malarial parasites?

Allison: Yes, granules have been shown in association with the Golgi apparatus of the malarial parasites and there are vesicles containing haemoglobin which is possibly being digested by various hydrolases. After chloroquin treatment, firstly there is fusion of vesicles with the formation of large structures, these also fuse with the cell wall (Warhurst and Hockley, 1967, *Nature* **214**, 935). This suggests that membranes become excessively sticky. Pigment then becomes discharged by a process rather like exocytosis. Secondly, large autophagic vacuoles form and it has been suggested that these are responsible for killing the parasite. This is supported by the observation that chloroquin kills parasites only during its intra-erythrocytic stages. Dr. Yielding favours the explanation that chloroquin binding to DNA interferes with DNA synthesis in accord with *in vitro* findings (Cohen and Yielding, 1965, *Proc. Nat. Acad. Sci. U.S.* **54**, 521).

Yielding: Dr. Fitch at Walter Reed has shown that certain malarial parasites which are resistant to chloroquin treatment do not have the same high affinity for chloroquin as that shown by sensitive parasites (Fitch, 1970, *Science* **169**, 289).

Allison: Yes, and in resistant parasites chloroquin does not produce the dramatic lysosomal response which I described.

Scriver: Is defective lysosome formation in the parasite a possible mechanism for the resistance to *P. falciparum* shown by subjects with sickle cell anaemia?

Allison: I have been interested in this possibility for some years. It is very difficult technically to get sickle cells to support multiplication of high concentrations of malarial parasites and the problem has not yet been resolved.

Spector: You mentioned that chlorpromazine is taken up by brain cells. Do you think that it might act by decreasing protein catabolism, or affect biologically functional proteins within membranes and thus influence the rate of protein synthesis within the brain?

Allison: Chlorpromazine interacts with numerous non-polar structures, including membranes in lysosomes, but I think it very unlikely that its tranquilising effect can be explained by its effect on lysosomes.

The Biochemistry of the Gangliosides in Relation to Inborn Disorders of Ganglioside Metabolism

N. B. MYANT

MRC Lipid Metabolism Unit,
Hammersmith Hospital, London, W12

Definition and Structure

The gangliosides belong to a group of complex lipids known as sphingolipids, so named because they contain a residue of sphingosine (from the Greek, Sphinx). Sphingosine was first isolated by Thudichum (1884), one of the founders of neurochemistry, who gave it this name 'in commemoration of the many enigmas which it presented to the enquirer'. Sphingosine is a C_{18} unsaturated amino alcohol with the structure shown in Fig. 1. In some gangliosides the sphingosine residue is replaced by dihydrosphingosine, the analogue of sphingosine in which the 4:5 double bond is saturated. A C_{20} homologue of sphingosine also appears in the brain gangliosides of human beings after birth, but is not detectable in the gangliosides of human foetal brain (Rosenberg, 1967). C_{20} sphingosine is not normally present in brain cerebrosides (but see Infantile Gaucher's disease).

All sphingolipids contain a ceramide residue. Ceramides are N-acyl sphingosines with the structure shown in Fig. 1, where

$$\mathrm{R{-}\overset{}{\underset{|}{C}}{=}O}$$

is a long-chain fatty acid attached in amide linkage to the amino group of sphingosine.

There is no generally accepted definition of a ganglioside, but for the purpose of this discussion I shall define a ganglioside as a sphingolipid containing a ceramide linked to a carbohydrate chain to which one or

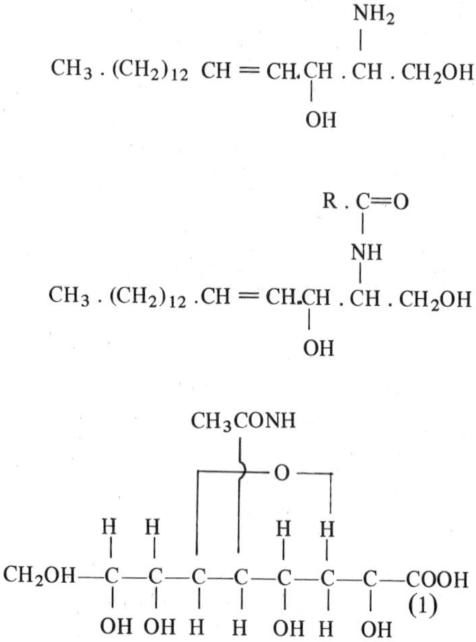

$$\begin{array}{c} \qquad\qquad\qquad\qquad NH_2 \\ \qquad\qquad\qquad\qquad | \\ CH_3 . (CH_2)_{12}\ CH = CH.CH . CH . CH_2OH \\ \quad | \\ \quad OH \end{array}$$

Sphingosine

$$\begin{array}{c} \qquad\qquad\qquad\qquad R . C{=}O \\ \qquad\qquad\qquad\qquad | \\ \qquad\qquad\qquad\qquad NH \\ \qquad\qquad\qquad\qquad | \\ CH_3 . (CH_2)_{12} .CH = CH.CH . CH . CH_2OH \\ \quad | \\ \quad OH \end{array}$$

Ceramide

$$\begin{array}{c} CH_3CONH \\ \text{—O—} \\ \begin{array}{cccccccc} & H & H & & & H & H & \\ & | & | & | & | & | & | & \\ CH_2OH— & C— & C— & C— & C— & C— & C— & C—COOH \\ & | & | & | & | & | & | & |\ \ (1) \\ & OH & OH & H & H & OH & H & OH \end{array} \end{array}$$

N-Acetyl Neuraminic acid (NANA)

(Structural formulae of sphingosine, ceramide and N-acetylneuraminic acid.)

Fig. 1.

more residues of a sialic acid are attached. In brain gangliosides the sialic acid is N-acetylneuraminic acid (NANA), shown in Fig. 1. The ceramide-carbohydrate chain has one or other of the following three structures:

(1) CER-GLU-GAL-GALNAc-GAL
(2) CER-GLU-GAL-GALNAc
(3) CER-GLU-GAL

where CER = ceramide, GLU = glucose, GAL = galactose and GALNAc = N-acetylgalactosamine. Each NANA group is attached either to a GAL or to another NANA. Thus, a representative ganglioside may be denoted in shorthand:

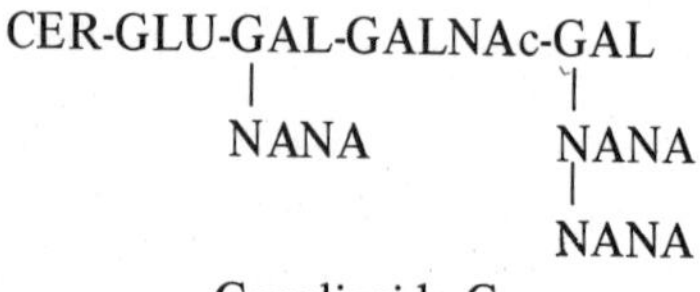

$$\begin{array}{l} \text{CER-GLU-GAL-GALNAc-GAL} \\ \qquad\quad\ \ |\qquad\qquad\quad\ \ | \\ \qquad\ \ \text{NANA}\qquad\ \ \text{NANA} \\ \qquad\qquad\qquad\qquad\quad\ \ | \\ \qquad\qquad\qquad\qquad\ \ \text{NANA} \end{array}$$

Ganglioside G_{T1}

More than 90% of the fatty acid of human brain ganglioside is stearic acid (C_{18}). In this respect, gangliosides differ from brain cerebrosides, in which a high proportion of the fatty acids are C_{20}, C_{22} and C_{24}.

Several codes for denoting gangliosides have been suggested. The one that is most convenient for spoken and written communication was devised by Svennerholm. In this system, G denotes ganglioside; subscripts M, D, T, Q and A denote mono-, di-, tri-, tetra- and asialo-gangliosides respectively; subscripts 1, 2 and 3 denote the ceramide-carbohydrate chains shown above. Thus, the trisialo-ganglioside shown above would be G_{T1}.

Table 10.1 shows some gangliosides of interest from the point of view of lipid storage diseases, with their structures denoted in shorthand and their symbols in the Svennerholm code. The closely related sphingolipid, globoside, is also included in this Table. Globoside,

TABLE 10.1

Structural Formulae (in shorthand), Codes and Trivial Names of Some Representative Gangliosides

	Chemical structure	*Code (Svennerholm)*	*Trivial name*
1.	CER-GLU-GAL(-NANA)-GALNAc-GAL(-NANA)	G_{D1}	
2.	CER-GLU-GAL(-NANA)-GALNAc-GAL	G_{M1}	
3.	CER-GLU-GAL(-NANA)-GALNAc	G_{M2}	Tay-Sachs ganglioside
4.	CER-GLU-GAL(-NANA)	G_{M3}	Haematoside
5.	CER-GLU-GAL-GALNAc	G_{A2}	
6.	CER-GLU-GAL	G_{A3}	Lactosyl ceramide
7.	CER-GLU-GAL-GAL-GALNAc		Globoside
8.	CER-GLU-GAL-GAL		Ceramide trihexoside

(Gangliosides G_{D1} and G_{M1} together account for about 90% of the total ganglioside content of mammalian brain. G_{M2} and G_{M3} are also present in normal brain, but in much smaller amounts. Globoside is a normal constituent of the membranes of red blood cells).

in which GLU and GALNAc are separated by two GAL residues, is a constituent of red-cell membranes and is probably the source of the glucocerebroside that accumulates in tissues outside the nervous system in Gaucher's disease in its adult form (but see Infantile Gaucher's disease).

Distribution and Function

Brain is by far the richest source of gangliosides in the body. Indeed, they were originally isolated from human brain tissue and were given the name by which they are generally known because of their prevalence in grey matter (Klenk, 1942). They have, however, been detected chromatographically in all tissues in which their presence has been sought, particularly in adrenal medulla, spleen and placenta (Svennerholm, 1970).

Very little is known about the physiological role of gangliosides. Their high concentration in brain suggests that they participate in some way in the specialised function of nerve tissue. McIlwain (1963) has suggested that brain gangliosides are concerned in the maintenance of the electrical excitability of brain tissue, not by acting as substrates for cell metabolism, but by providing negative charges which interact with histones and other basic constituents of nerve cells. The study of the cellular and subcellular distribution of brain gangliosides has not provided any very promising clues to their function. They are present in highest concentration in the neuronal cells of grey matter, but traces have also been found in myelin (Eichberg *et al.*, 1964; Suzuki *et al.*, 1968). Their concentration in microsomes is higher than that in other subcellular fractions, but they are detectable in mitochondria and in the cytosol (Eichberg *et al.*, 1964). The possibility that brain gangliosides provide the physical basis for information storage is an intriguing one. In order to perform such a rôle they would presumably have to exist in the intact cell in a state of ordered polymerisation. Although gangliosides readily form aggregates of high molecular weight in aqueous media, these aggregates are now thought to be micelles, in which the arrangement of the molecules is randomised, rather than true polymers.

Synthesis

The elaboration of a ganglioside molecule consists essentially in the formation of ceramide from sphingosine, followed by the stepwise addition of monosaccharide units to a priming unit of ceramide. The biosynthesis of sphingosine has been elucidated by Brady and his colleagues (Brady and Koval, 1958; Brady *et al.*, 1958). Synthesis in brain takes place by a condensation between palmitaldehyde and serine, with loss of CO_2 from the carboxyl group of serine. This rather complex reaction, which requires Mn^{2+}, pyridoxal phosphate and a hydrogen acceptor, may be simplified to:

$$\underset{\text{Palmitaldehyde}}{CH_3(CH_2)_{12}\,.\,CH_2\,.\,CH_2\,.\,CHO} + \underset{\text{Serine}}{COOH\,.\,CH(NH_2)CH_2OH} \quad (1)$$

$$\rightarrow CH_3(CH_2)_{12}\,.CH{=}CH.\,\underset{\substack{|\\ OH}}{CH}\,.\,\overset{\substack{NH_2\\ |}}{CH}\,.\,CH_2OH + CO_2 + 2H$$

Sphingosine

All the enzymes required for this synthesis are present in the brains of developing animals. Ceramide is synthesised by the condensation of the fatty acid residue of fatty acyl CoA with the amino group of sphingosine:

$$\text{Sphingosine} + \text{Fatty Acyl CoA} \rightarrow \text{Ceramide} + \text{CoA} \quad (2)$$

This reaction is catalysed by an enzyme present in liver and brain (Sribney, 1966).

The addition of the carbohydrate chain to the ceramide primer has been studied in some detail by Kaufman *et al.*, 1967, 1968) and by Basu *et al.* (1968). GLU, GAL, GALNAc and NANA are transferred from their nucleotide derivatives to the growing lipid acceptor, each reaction being catalysed by a specific transferase. The first six of these reactions, resulting in the formation of G_{D1}, are as follows:

$$\text{CER} + \text{UDP-GLU} \rightarrow \text{CER-GLU} + \text{UDP} \quad (3)$$

$$\text{CER-GLU} + \text{UDP-GAL} \rightarrow \text{CER-GLU-GAL} + \text{UDP} \quad (4)$$

$$\text{CER-GLU-GAL} + \text{CMP-NANA} \rightarrow \text{CER-GLU-}\underset{\substack{|\\ \text{NANA}}}{\text{GAL}} + \text{CMP} \quad (5)$$

$$\text{CER-GLU-}\underset{\substack{|\\ \text{NANA}}}{\text{GAL}} + \text{UDP-GALNAc} \rightarrow \text{CER-GLU-}\underset{\substack{|\\ \text{NANA}}}{\text{GAL}}\text{-GALNAc} + \text{UDP} \quad (6)$$

```
CER-GLU-GAL-GALNAc + UDP-GAL                                        (7)
        |
       NANA
       → CER-GLU-GAL-GALNAc-GAL + UDP
                 |
                NANA

CER-GLU-GAL-GALNAc-GAL + CMP-NANA                                   (8)
        |
       NANA
       → CER-GLU-GAL-GALNAc-GAL + CMP
                 |            |
                NANA         NANA
```

Brain contains two galactosyl transferases concerned in ganglioside synthesis, one catalysing reaction (4) and the other catalysing the introduction of the terminal GAL of the carbohydrate chain to form ganglioside G_{M1} (reaction (7)). There appear to be at least two NANA transferases in brain, one for reaction (5) and another for reaction (8). There may be yet other NANA transferases catalysing the addition of the third and fourth NANA groups to form ganglioside G_{Q1}, but the evidence for the presence of these enzymes in brain is incomplete. From what is known of the specificities of the enzymes catalysing the synthesis of gangliosides, it seems likely that they act in a concerted manner, the preferred substrate of each enzyme being the product of the previous reaction in the biosynthetic sequence. Basu *et al.* (1968) suggest the term 'multiglycosyltransferase system' for the group of enzymes carrying out these reactions.

Breakdown

Current views on the breakdown of brain gangliosides are based partly on the study of enzymes obtained from normal tissues, and partly on the identification of metabolites that accumulate in the bodies of patients with inborn errors of ganglioside metabolism. At least five enzymes concerned in the complete breakdown of gangliosides have been obtained in various degrees of purity from brain. All have pH optima on the acid side of neutrality and their subcellular distribution suggests that all of them are lysosomal. The five enzymes for which there is direct evidence are listed below:

Neuraminidase (EC 3.2.1.18), which removes NANA from the carbohydrate chain of a ganglioside;

β-Galactosidase (EC 3.2.1.23), which removes both GAL residues.

(Indirect evidence for the existence of more than one β-galactosidase in brain is discussed below);

β-N-Acetylhexosaminidase (β-hexosaminidase) (EC 3.2.1.30), which removes GALNAc from G_{M2} ganglioside;

β-Glucosidase (EC 3.2.1.21), which hydrolyses CER-GLU to CER and GLU;

Ceramidase, which splits CER into sphingosine and a fatty acid.

The substrate specificity of these enzymes is such that they probably act sequentially according to a scheme proposed by Gatt (1967). In this scheme (Table 10.2), G_{M1} is formed from higher gangliosides by the removal of all the NANA residues except the one attached to the internal GAL. The terminal GAL and the GALNAc residue are then removed successively, giving rise to G_{M3}, from which the NANA is removed to give G_{A3}; the further breakdown is then as shown in Table 10.2. This is probably the major pathway for ganglioside breakdown in brain. There may, however, be alternative minor pathways in which the last NANA is removed before loss of the GALNAc residue, since the NANA-free 'gangliosides' G_{A1} and G_{A2} accumulate in the brain when the breakdown, respectively, of G_{M1} and G_{M2} is blocked by an inborn error of metabolism (see below). Evidently, steric hindrance to the removal of the last NANA by the adjacent GALNAc residue is not complete. Gatt (1967) has pointed out that several of the enzymes catalysing the breakdown of gangliosides are localised in the same subcellular particles (probably the lysosomes), that they are not solubilised by sonication, and that they have similar values for pH optimum, K_m and v_{max}. He suggests that they may be linked together to form a multienzyme complex bound to a membrane.

Inborn Errors of Ganglioside Metabolism

Ganglioside storage diseases are discussed in detail in this symposium by Dr. Raine and will only be considered here in so far as they throw light on the normal pathways of ganglioside metabolism.

Excessive accumulation of a ganglioside could be caused by a defect in the biosynthetic pathway in which it occurs. It has been suggested, for example, that G_{M2} accumulates in Tay-Sachs disease because of a failure to convert G_{M2} into G_{M1} by addition of the terminal GAL (Basu *et al.*, 1965) (see Table 10.2). The most recent evidence suggests, however, that in all those diseases in which there is accumulation of a ganglioside or related sphingolipid, the underlying abnormality is deficiency of a hydrolytic enzyme catalysing a step in a catabolic pathway. The gangliosidoses thus appear to belong to a larger group of

TABLE 10.2

Probable Pathways for the Breakdown of Higher Brain Gangliosides (modified from Gatt, 1967) and of Globoside

Step		Chemical structure	Code	Storage disease
1.		CER-GLU-GAL-GALNAc-GAL NANA NANA NANA NANA [Neuraminidase]	G_{Q1}	
2.	G_{A1} <···	CER-GLU-GAL-GALNAc-GAL NANA [β-Galactosidase]	G_{M1}	G_{M1} gangliosidosis (pseudo-Hurler)
3.	G_{A2} <···	CER-GLU-GAL-GALNAc NANA [β-Hexosaminidase]	G_{M2}	Tay-Sachs disease
4.		CER-GLU-GAL NANA [Neuraminidase]	G_{M3}	Patient of Pilz *et al.* (1966)
5.	·····>	CER-GLU-GAL [β-Galactosidase]	G_{A3}	Patient of Pilz *et al.* (1966)
6.		CER-GLU [β-Glucosidase]		Infantile Gaucher's disease
7.		CER Ceramidase → Sphingosine + fatty acid		
	?	CER-GLU-GAL-GAL-GALNAc (Globoside) [β-Hexosaminidase]		Tay-Sachs with visceral involvement
		*CER-GLU-GAL-GAL [α-Galactosidase]		Fabry's disease
		CER-GLU-GAL		
		CER-GLU [β-Glucosidase]		Adult Gaucher's disease
		CER		

* If the GAL-GAL linkage in globoside has the β-configuration, then the ceramide trihexoside that accumulates in Fabry's disease could not be derived from globoside.

sphingolipidoses, including Niemann-Pick disease, Fabry's disease, Gaucher's disease and metachromatic leucodystrophy, in all of which there is an inherited deficiency of a specific lysosomal acid hydrolase.

With the possible exception of an abnormal accumulation of a disialoganglioside (G_{D1}), reported in the brain of one infant (Volk *et al.*, 1964), there are no known disorders of ganglioside metabolism attributable to a deficiency of neuraminidase. There are, however, several that appear to be due to deficiency of a β-galactosidase or of a β-hexosaminidase.

β-Galactosidase Deficiency

In the familial disease known as generalised gangliosidosis (G_{M1} gangliosidosis), G_{M1} and the NANA-free 'ganglioside', G_{A1}, accumulate in the brain and viscera. This disease has been shown by Okada and O'Brien (1968) to be due to a generalised deficiency of the β-galactosidase that catalyses the removal of the terminal GAL from G_{M1} and G_{A1} (Table 10.2, Step 2). The possibility that there are two forms of this disease, associated with deficiencies of different isoenzymes of β-galactosidase, is discussed by Dr. Raine in this Symposium.

Pilz *et al.* (1966) have reported increased amounts of G_{M3} and G_{A3} (lactosyl ceramide) in the brain of an infant who died from an unusual form of gangliosidosis. This finding suggests that in this infant there was a metabolic block at Step 5 (Table 10.2) due to an inborn deficiency of a β-galactosidase. If this was so, the missing enzyme must be specific for the internal GAL of the carbohydrate chain, because there was no accumulation of G_{M1} in the infant's brain.

In Fabry's disease (glycolipid lipidosis) the ceramide trihexoside CER-GLU-GAL-GAL accumulates in the central nervous system, skin, kidneys and other organs. Brady *et al.* (1967) have shown that human intestinal wall contains a galactosidase that catalyses the hydrolytic removal of the terminal GAL of this trihexoside to give lactosyl ceramide and galactose:

$$\underset{\text{Ceramide trihexoside}}{\text{CER-GLU-GAL-GAL}} \rightarrow \underset{\text{Lactosyl ceramide}}{\text{CER-GLU-GAL}} + \text{GAL} \qquad (9)$$

In patients with Fabry's disease this enzyme is absent from the intestine (Brady *et al.*, 1967). These observations suggest that the accumulation of ceramide trihexoside is due to a generalised absence of a galactosidase specific for the terminal GAL of ceramide trihexoside. Since G_{M1} does not accumulate in the tissues in Fabry's disease, the missing enzyme must be different from the one that catalyses the removal of the terminal GAL from G_{M1}. In confirmation of this, Kint (1970) has shown that α-galactosidase (EC 3.2.1.22), an enzyme present in normal leucocytes, is absent from the leucocytes of patients suffering from Fabry's disease. If the enzyme that normally hydrolyses ceramide trihexoside is α-galactosidase, and not β-galactosidase, then the terminal GAL must have the α-configuration. In the five patients studied by Brady *et al.* (1967) the activity of lactosyl ceramidase, the enzyme catalysing the hydrolysis of the product of reaction (9), was subnormal. Brady (1968) has made the interesting suggestion that this may have been due to lack of induction of this enzyme owing to the absence of its substrate (lactosyl ceramide). The ceramide trihexoside that accumulates in the tissues in Fabry's disease may be derived from red-cell globoside, since the trihexoside and globoside have an identical fatty acid composition (Sweeley and Klionsky, 1963) and both have the same CER-GLU-GAL-GAL sequence.

β-Hexosaminidase Deficiency

In patients with the more usual form of Tay-Sachs disease (infantile amaurotic family idiocy), ganglioside G_{M2} and its NANA-free derivative, G_{A2}, accumulate in the brain. Both these compounds have the terminal GALNAc residue which is removed by the β-hexosaminidase present in normal brain (see Table 10.2). This suggests that Tay-Sachs disease is due to a deficiency of β-hexosaminidase. Recent work supports this suggestion, but several rather puzzling questions have yet to be answered. In human tissues, there are two iso-enzymic forms of β-hexosaminidase associated with lysosomes (Robinson and Stirling, 1968). These two iso-enzymes can be separated by electrophoresis and are referred to as the A and B forms. Okada and O'Brien (1969), using an artificial substrate for assaying these enzymes, have shown that in the usual form of Tay-Sachs disease, β-hexosaminidase A is absent from brain, leucocytes, liver and other tissues, whereas the activity of β-hexosaminidase B is considerably increased. If it could be shown that G_{M2} and G_{A2} are substrates for β-hexosaminidase A but not for the B enzyme, then it would be hard to avoid the conclusion that the underlying defect in Tay-Sachs' disease is an inherited deficiency of β-hexosaminidase A. However, according to Jatzkewitz, *et al.* (1970), the GALNAc residue of G_{M2} is removed by both enzymes. This makes it difficult to understand why G_{M2} is not metabolised by the β-hexosaminidase B present in the brains of Tay-Sachs' patients.

A possible explanation for this anomaly is that enzyme specificity may, in some cases, be influenced by the physical state of the substrate.

The action of lysosomal hydrolases on gangliosides *in vitro* is markedly influenced by the addition of detergents to the incubation mixture (Gatt, 1967), suggesting that the micellar structure of the substrate influences its susceptibility to enzymic attack. It is possible that ganglioside G_{M2} in its natural state can act as substrate only for the A form of hexosaminidase, and that when the ganglioside is extracted from tissues and tested in the presence of detergent, its physical state is changed in a way such as to make it susceptible to hydrolysis by both forms of the enzyme.

A further difficulty in explaining Tay-Sachs disease in terms of a deficiency of hexosaminidase A arises from a recent observation of Sandhoff (1969), who found that both iso-enzymes were present in normal amounts in the brain of an infant who had died from this disease.

Sandhoff *et al.* (1968) have described an unusual form of Tay-Sachs disease in which, in addition to storage of G_{M2} and G_{A2} in the brain, globoside accumulates in the kidneys and other viscera. All three of these compounds contain a terminal GALNAc residue, suggesting a deficiency of β-hexosaminidase. In agreement with this, patients suffering from this disease have been shown to lack both the A and the B forms of β-hexosaminidase in brain (Sandhoff, 1969; O'Brien, 1969). Since globoside does not accumulate in the usual form of Tay-Sachs disease, in which the A enzyme is deficient, this finding suggests that the GALNAc residue of globoside is removed by hexosaminidase B but not by the A form of the enzyme.

Infantile Gaucher's Disease

Although not a true gangliosidosis, the rare infantile form of Gaucher's disease must be mentioned here because of its relevance to the metabolism of brain gangliosides. In the more usual ('adult') form of this disease the central nervous system is not affected. The lipid that accumulates in the tissues is a glucocerebroside (CER-GLU), resembling the globoside of red blood cells in its fatty acid composition and in its lack of the C_{20} homologue of sphingosine (Suomi and Agranoff, 1965). Because of these similarities it is generally assumed that the storage lipid in adult Gaucher's disease is derived from globoside.

In the infantile form of the disease, the central nervous system (in addition to the liver, spleen and haemopoietic system) is affected, the patient usually dying in infancy from bulbar palsy. Svennerholm (1967) examined the cerebral grey matter of three infants who had died from this disease and found that galactocerebroside (CER-GAL), the predominant cerebroside of normal brain, was largely replaced by glucocerebroside. However, the fatty acid composition of the brain glucocerebroside resembled that of brain ganglioside, and differed from that of globoside. Moreover, the brain cerebroside had the same C_{20} sphingosine content as brain ganglioside. Svennerholm therefore suggested that the cerebroside stored in the brains of these infants was derived from brain gangliosides and that its accumulation was due to an inborn deficiency of β-glucosidase in brain.

This raises the question as to why the brain is not affected in the adult form of Gaucher's disease, in which there is an inherited deficiency of β-glucosidase activity in spleen and other non-neural tissues (Brady, 1968). The existence of the two genetically distinct forms of this disease could be explained by postulating two different β-glucosidases catalysing the hydrolysis of CER-GLU cerebrosides, one in brain and the other a 'splenic' enzyme. On this hypothesis, adult Gaucher's disease is due to an inherited deficiency of the splenic enzyme alone and the infantile form is due to a double genetic defect causing a deficiency of both enzymes. An alternative explanation is that there is a single β-glucosidase for all glucocerebrosides in the body and that the genetic defect in adult Gaucher's disease is such as to lead to a moderate loss of enzyme activity, whereas that in the infantile form leads to a more complete loss, perhaps because it affects a region of the enzyme very close to the active site. If a moderate fall in enzyme activity affects the breakdown of glucocerebrosides in non-neural tissue, but not of those in brain, then the existence of two forms of Gaucher's disease could be explained on the basis of two levels of genetically determined enzyme deficiency. In favour of this explanation, Brady *et al.* (1966) found that the activity of glucocerebrosidase in spleens from patients with the adult form of Gaucher's disease was 15% of that in normal spleen, whereas its activity in spleens from patients with the infantile form was negligible or not detectable.

Concluding Remarks

Each of the disorders of sphingolipid metabolism referred to in this chapter is due to a defect in a catabolic pathway, resulting in the pathological accumulation of a normal intermediate which cannot be removed from the body. There is already evidence for the existence of at least seven inborn disorders of the breakdown of gangliosides, including the globosides, and recent work on iso-enzymes of β-galactosidase and β-hexosaminidase (O'Brien, 1969) suggests that the number of distinct genetic abnormalities affecting ganglioside breakdown will eventually turn out to be considerably greater than this.

In every case in which an enzyme defect has been shown to be the cause of the metabolic block, the enzyme appears to be an acid hydrolase associated with the lysosomes. This is true not only of the gangliosidoses, but also of other sphingolipidoses.

In view of the large number of enzymes required for the synthesis of a ganglioside, it may seem surprising that no inborn disorder of ganglioside metabolism has yet been shown to be due to deficiency of an enzyme catalysing a biosynthetic step. This can hardly be because the genes controlling the synthesis of these enzymes do not mutate. A more likely explanation is that an inborn inability to form gangliosides is not compatible with the growth and development of the embryo beyond a very early stage of intra-uterine life.

Another question worth considering here is the age at which the clinical signs of a sphingolipidosis first appear, since this must be an important factor in determining the degree of mental retardation associated with the disease. In some sphingolipidoses, clinical signs appear in the first few months of life, or may even be present at birth; in others they do not appear until much later on in life. Jatzkewitz (1969) has discussed some of the factors that determine the rate at which intermediates would be expected to accumulate when there is a block in the pathway for catabolism of a sphingolipid. Clearly, the greater the rate of turnover under normal conditions, the more rapidly will an intermediate accumulate if its breakdown is completely blocked. Brain gangliosides are in a state of relatively rapid turnover before birth. Hence, a complete block in their breakdown could lead to accumulation of gangliosides in the brain *in utero.* In keeping with this, in one form of Tay-Sachs disease (congenital amaurotic family idiocy) signs of involvement of the central nervous system are present at birth and the retardation of mental development in the neonatal period is extreme. Another factor that must influence the rate at which sphingolipid intermediates accumulate is the extent to which the gene mutation affects the activity of the corresponding enzyme. A mutation which resulted in the formation of a completely inactive enzyme would be expected to lead to a more rapid accumulation of its substrate, and hence to the earlier appearance of clinical symptoms, than one which resulted in only a partial loss of enzyme activity. As we have already seen, this seems to be the explanation of the existence of the two forms of Gaucher's disease. It is not yet known whether differences in enzyme activity can explain differences in the severity and age at onset of other sphingolipidoses. However, there is evidence to suggest that two types of Tay-Sachs disease, one with early and the other with late onset, are due, respectively, to complete and partial deficiency of β-hexosaminidase A (O'Brien, 1969).

REFERENCES

Basu, S., Kaufman, B. and Roseman, S. (1965). *J. biol. Chem.* **240,** PC 4115.
Basu, S., Kaufman, B. and Roseman, S. (1968). *J. biol. Chem.* **243,** 5802.
Brady, R. O. (1968). *Advanc. clin. Chem.* **11,** 1.
Brady, R. O., Formica, J. V. and Koval, G. J. (1958). *J. biol. Chem.* **233,** 1072.
Brady, R. O., Gal, A. E., Bradley, R. M. and Martensson, E. (1967). *J. biol. Chem.* **242,** 1021.
Brady, R. O., Gal, A. E., Bradley, R. M., Martensson, E., Warshaw, A. L. and Laster, L. (1967). *New Engl. J. Med.* **276,** 1163.
Brady, R. O., Kanfer, J. N., Shapiro, D. and Bradley, R. M. (1966). *J. clin. Invest.* **45,** 1112.
Brady, R. O. and Koval, G. J. (1958). *J. biol. Chem.* **233,** 26.
Eichberg, J., Jr., Whittaker, V. P. and Dawson, R. M. C. (1964). *Biochem. J.* **92,** 91.
Gatt, S. (1967). *Biochim. biophys. Acta* **137,** 192.
Jatzkewitz, H. (1969). In 'Some Inherited Disorders of Brain and Muscle' (eds J. D. Allan and D. N. Raine), pp. 114-129. Proceedings of the Fifth Symposium of The Society for the Study of Inborn Errors of Metabolism. Edinburgh: E. and S. Livingstone.
Jatzkewitz, H., Mehl, E. and Sandhoff, K. (1970). *Biochem. J.* **117,** 6P.
Kaufman, B., Basu, S. and Roseman, S. (1967), In 'Inborn Disorders of Sphingolipid Metabolism' (eds S. M. Aronson, and B. W. Volk), pp. 193-213. Proceedings of the Third International Symposium on The Cerebral Sphingolipidoses. Oxford: Pergamon Press.
Kaufman, B., Basu, S. and Roseman, S. (1968). *J. biol. Chem.* **243,** 5804.
Kint, J. A. (1970). *Science* **167,** 1268.
Klenk, E. (1942). *Ber. dtsch. chem. Ges.* **75,** 1632.
McIlwain, H. (1963). 'Chemical Exploration of the Brain. A Study of Cerebral Excitability and Ion Movement'. Amsterdam: Elsevier.
O'Brien, J. S. (1969). *Lancet* **2** 805.
Okada, S. and O'Brien, J. S. (1968). *Science* **160,** 1002.
Okada, S. and O'Brien, J. S. (1969). *Science* **165,** 698.
Pilz, H., Sandhoff, K. and Jatzkewitz, H. (1966). *J. Neurochem.* **13,** 1273.
Robinson, D. and Stirling, J. L. (1968). *Biochem. J.* **107,** 321.
Rosenberg, A. (1967). In 'Inborn Disorders of Sphingolipid Metabolism' (eds S. M. Aronson and B. W. Volk), pp. 267-272. Proceedings of the Third International Symposium on The Cerebral Sphingolipidoses. Oxford: Pergamon Press.
Sandhoff, K. (1969). *FEBS Letters* **4,** 351.
Sandhoff, K., Andreae, U. and Jatzkewitz, H. (1968). *Life Sci.* **7,** 283.
Sribney, M (1966). *Biochim. biophys. Acta* **125,** 542.
Suomi, W. D. and Agranoff, B. W. (1965). *J. Lipid Res.* **6,** 211.
Suzuki, K., Poduslo, J. F. and Poduslo, S. E. (1968). *Biochim. biophys. Acta* **152,** 576.
Svennerholm, L. (1967). In 'Inborn Disorders of Sphingolipid Metabolism' (eds S. M. Aronson and B. W. Volk), pp. 169-186. Proceedings of the Third International Symposium on The Cerebral Sphingolipidoses. Oxford: Pergamon Press.
Svennerholm, L. (1970). In 'Comprehensive Biochemistry' (eds M. Florkin and E. H. Stotz), pp. 201-227. Amsterdam: Elsevier.
Sweeley, C. C. and Klionsky, B. (1963). *J. biol. Chem.* **238,** PC3148.

Thudichum, J. L. W. (1884). 'A Treatise on the Chemical Constitution of The Brain', p. 105. London: Baillière, Tindall and Cox.
Volk, B. W., Wallace, B. J., Schneck, L. and Saifer, A. (1964). *Arch. Path.* **78**, 483.

DISCUSSION

Stern: I think one should not forget that in a number of inborn errors of metabolism there may be an appreciable residual activity of the deficient enzyme, for example the deficiency of hexosaminidase A is less severe in late onset G_{M2} gangliosidosis than classical Tay-Sachs disease (O'Brien, 1969, *Lancet* **2**, 805).
Scriver: Yes. This reminds me of Pinsky's work on cultured fibroblasts from subjects with generalised G_{M1} gangliosidosis (types 1 and 2). In type 1 there is virtually no β-galactosidase activity. This type has a very severe course resulting in early death. In type 2 however in which the fibroblasts have 2-3% of the normal total activity the clinical course is milder.
Raine: O'Brien has stated that in starch gel electrophoresis of liver in type 2 G_{M1} gangliosidosis, one of the three bands of β-galactosidase is present, while in type 1 all three bands are absent. It would be interesting to know if these differences could be demonstrated in fibroblasts.
McIlwain: Some time ago we studied the function of gangliosides and concluded that they could act by contributing to the ionic conditions on the outside of neural cells. The exchange of sodium and potassium ions which occurs during nerve impulses can be blocked by a number of basic peptides and histones; but this blocking effect can be reversed and normal ionic movement restored by adding ganglioside agents.
Robinson: We have evidence that the two types of hexosaminidases (A and B) from human spleens are closely related to each other. Treatment of the A types with neuraminidase changes it to a form which closely resembles the B type. Moreover the A type of hexosaminidase derived from human brain is converted to the B type simply on standing. The reports by Sandhoff *et al.* (1968, *Pathol. Europaea* **3**, 278), and O'Brien *et al.* (1970, *New Eng. J. Med.* **283**, 15) that in some cases of Tay-Sachs disease there may be normal or elevated amounts (presumably of the B enzyme) should be interpreted with these observations in mind. One may postulate that there may be only one gene for both A and B enzymes, and possibly in Tay-Sachs disease there may be a defect in conversion of the B form into the A.

Myant: A mutation involving a common gene may be the explanation in cases where there is a deficiency of both A and B enzymes.

Yielding: In some cases it has been shown that certain procedures, e.g. detergent treatment can alter enzyme specificity.

Robinson: Sandhoff (1969, *FEBS Letters* **4**, 351) has shown that when Tay-Sachs hexosaminidase is fractionated, the B form may be present but there may be a large peak of inert protein in the position where the A form is usually present. Possibly this latter situation exists in a type of Tay-Sachs disease in which a localised mutation may lead to a few incorrect amino acids in the active centre, so that inactive enzyme protein is produced.

One should consider whether deficiency of the A form of hexosaminidase may not be a feature of a more widespread mucopolysaccharide abnormality. The A form may just be the B form with a piece of mucoid material attached to it. The B form is very basic and will readily interact with mucopolysaccharides such as heparin producing a complex which has a different electrophoretic mobility to the pure B form but which resembles that of the A form. Perhaps, therefore, we should examine other glycoproteins in patients with Tay-Sachs disease.

Jepson: Other hydrolytic enzymes which can be converted into other forms by treatment with neuraminidase are the various alkaline phosphatases (Smith *et al.*, 1968, *Clin. Chim. Acta* **19**, 499).

Muir: To come back to partial enzyme activity, β-galactosidase deficiency in Hurler-Hunter syndrome is only partial and there may be up to 20% of the normal activity present.

Myant: Some discrepancies in observations of β-galactosidase activity may arise from the use of synthetic rather than natural substrates. In some studies the substrate is G_{M2} isolated from viscera of patients; in others, β-methyl umbelliferone derivatives are used.

Benson: Deficiency of β-galactosidase activity (using both o-nitrophenyl-β-D-galactopyranoside and ^{3}H-G_{M1} as substrates) has also been reported in cultured fibroblasts derived from skin and bone marrow from subjects with G_{M1} (type 1) gangliosidosis (Sloan *et al.*, 1969, *Pediat. Res.* **3**, 532).

Laboratory Diagnosis of Some Sphingolipidoses

D. N. RAINE

Biochemistry Department, The Children's Hospital, Birmingham B16 8ET

The term 'ganglioside' was given by Klenk to the substance he isolated from the brain of a patient with infantile amaurotic idiocy because it occurred mainly in the ganglion cells of the central nervous system. It has given rise to a series of compounds of sphingosine combined with a fatty acid (the two being known as ceramide) linked with one or more monosaccharides and with one or more sialic acid residues. In practice these compounds are found to occur with the corresponding sialic acid-free substance. Diseases characterised by the accumulation of these gangliosides and their asialo derivatives are now known as gangliosidoses and are designated by the letter G followed by a letter (A, none; M, mono-; D, di- and T, tri-) indicating the number of sialic acid residues. Futher distinction is given by a number and sometimes by a further lower case letter. The metabolic relationships of all of these are shown in Fig. 1.

This new nomenclature is replacing the former eponymous classification of amaurotic family idiocy and related disorders and to some extent crosses the former divisions. The new nomenclature and classification is justifying itself, however, since it is based on more fundamental aetiological grounds and has given rise to much more precise methods of diagnosing the several members of this group of diseases. With the exception of juvenile amaurotic family idiocy (Spielmeyer-Vogt disease) the cause of which is still obscure, the classification in terms of age of onset (congenital, Norman-Wood; infantile, Tay-Sachs; late infantile, Jansky-Bielschowsky; and adult, Kufs) is not very helpful.

It is convenient at the same time to consider diseases associated with the storage of some closely related compounds, without sialic acid and derived from globoside, and in addition to include the disorders of sulphatide metabolism, as these too are closely related chemically.

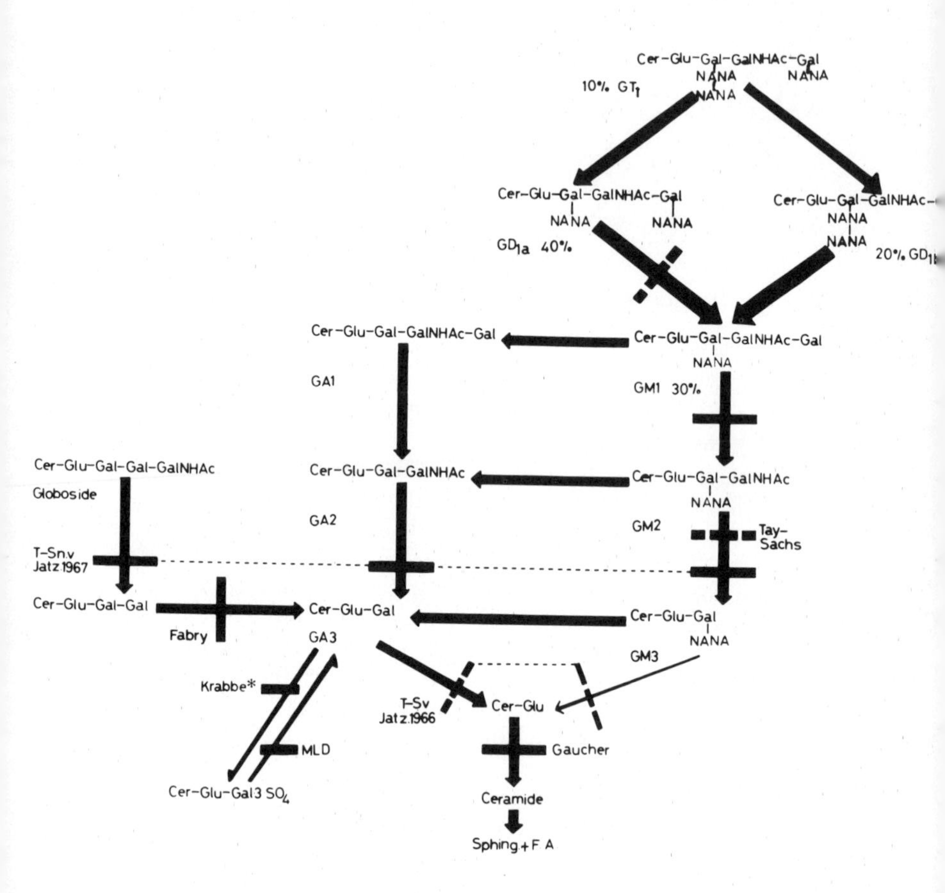

Fig. 1. Metabolic relationships of gangliosides and related compounds.

Cer = ceramide; Glu = glucose; Gal = galactose; GalNHAc = N-acetyl galactosamine; NANA = sialic acid.

The proportion of the first four gangliosides in normal brain is indicated. Solid bars represent established enzyme deficiencies; broken bars represent presumed enzyme deficiencies.

* (The defect in Krabbe's leucodystrophy has been questioned by Suzuki and Suzuki (1970), who now consider it a β-galactosidase deficiency.)

Three of these additional diseases, globoside storage disease, metachromatic leucodystrophy and Krabbe leucodystrophy, involve the nervous system but the fourth, Fabry's disease has entirely different clinical manifestations. The clinical and biochemical relationships of all of these diseases has been extensively reviewed (Raine, 1969; Raine, 1970). It is already evident that the complete story will be more complicated than is indicated by Fig. 1 as it is not yet apparent how the mild form of G_{M1} gangliosidosis (Wolfe *et al.*, 1970) will fit into this scheme. There may also be variants of metachromatic leucodystrophy (Griffiths *et al.*, 1970) and the underlying cause of these is not yet elucidated. Finally, it is by no means clear why some patients with the same apparent defect have their clinical onset at such different ages (Jatzkewitz, 1969).

There are several disorders that can be discussed with some confidence at the present time, however, and these include G_{M1} gangliosidosis (generalised gangliosidosis, Tay-Sachs with visceral involvement, pseudo-Hurler's disease), G_{M2} gangliosidosis (infantile or true Tay-Sachs disease), Gaucher's disease (the classical adult variety though not yet the congenital form), globoside storage disease (Tay-Sachs with hexosaminidase deficiency of Sandhoff *et al.*, 1968, and G_{M2} type II of O'Brien, 1969b), Fabry's disease (angiokeratoma corporis diffusum) and metachromatic leucodystrophy.

Analytical Approaches to Diagnosis

All of these diseases are inherited deficiencies of an enzyme involved in complex lipid metabolism. They may be investigated, therefore, by assaying the enzyme or by seeking evidence for an abnormal accumulation of normal lipid or of a lipid not normally detectable; both the lipid and the enzyme may be approached in different ways.

The lipid may accumulate in microscopically recognisable aggregates, this certainly occurs in the brain but it may also be seen in bone marrow cells. The same lipid may be extracted from tissue taken by biopsy from the brain, liver, spleen or kidney or it may be detectable after extracting it from the cells in the deposit obtained by centrifuging freshly passed urine. If there is sufficient material to subject it to thin-layer chromatography, the characteristic lipids, and hence the disease, can be recognised with considerable certainty.

Many of the diseases being considered were first characterised by establishing the deficiency of the enzyme concerned in a particular metabolic step, using brain or other tissue obtained after death. These same enzymes can be assayed in more accessible tissues and fluids during life and examples will be referred to later. In some instances, however, an assay system using an unnatural substrate has proved more convenient and, while this can still be useful in practice, it is necessary to consider the relationship between the natural enzyme whose deficiency is the cause of the disease and the enzyme activity whose assay leads to its diagnosis.

Thus, metachromatic leucodystrophy is due to a deficiency of brain cerebroside sulphate (sulphatide) sulphatase and when radioactively labelled sulphatide is used as substrate, the enzyme activity of the brain of affected subjects is found to be low. Hydrolysis of the synthetic p-nitrocatechol sulphate by brain tissue is also low but when the properties of this arylsulphatase are investigated three different forms (A, B and C) of the enzyme can be recognised. Although the activity of arylsulphatase-A correlates most closely with sulphatide sulphatase, the synthetic substrate can be hydrolysed under circumstances in which hydrolysis of the natural substrate does not occur. Indeed, the two enzyme activities develop at a different rate in the growing brain (Jatzkewitz, 1969). As long as the reasons for this difference remain obscure, diagnoses based on arylsulphatase activity alone must continue to be interpreted with caution.

Materials Available for Analysis

Formerly, clinical suspicion of most of the diseases being considered required study of brain tissue, at either biopsy or necropsy, for the diagnosis to be confirmed. In some instances peripheral nerve conduction velocity, peripheral nerve biopsy, rectal biopsy or examination of cells obtained from urinary sediment or from bone marrow biopsy could further the diagnosis. Now, however, chemical techniques, such as thin-layer chromatography and the more sophisticated enzyme assays, can be applied to a variety of tissues and fluids, to give an absolute diagnosis, not only in life with minimal interference with the patient, but also, by examining cells from amniotic fluid, *in utero.*

Circulating leucocytes contain the enzyme involved in all six diseases being considered and a deficiency of the enzyme has been demonstrated in the leucocytes of patients with each condition.

Urine enzyme activity is often high enough to be measured easily and in some instances the enzyme involved has been shown to be excreted in very low concentration by patients with the corresponding disease. Although this may be a useful confirmatory test, the variation associated with any urinary study makes it less precise than some other investigations.

Plasma enzyme activity has also proved useful in diagnosis, for when

TABLE 11.1

Methods so far established for the Diagnosis of some Sphingolipidoses

Disease	GM_1 ganglio-sidosis	GM_2 ganglio-sidosis	Gaucher's disease	Globoside storage disease	Fabry's disease	Metachromatic leucodys-trophy
Specific enzyme	β-galactosidase	galacto-saminidase A	glucos-idase	galacto-saminidase A and B	α-galacto-sidase	arylsulphatase-A
Circulating LEUCOCYTE enzyme deficiency	1	3, 5	6	1	7, 8, 17	1, 10 11, 12
Low enzyme activity in URINE	2					9
Low enzyme activity in PLASMA		3, 4			14	
Low enzyme activity in FIBROBLASTS		3, 4		1		12
Abnormal LIPID in URINE sediment				1	15	18
Low enzyme activity in AMNIOTIC fluid CELLS		16				13

1. Raine, 1969b and unpublished.
2. Thomas, 1969.
3. Okada and O'Brien, 1969.
4. O'Brien, 1969a.
5. O'Brien, 1969b.
6. Kampine, Brady, Kanfer, Feld and Shapiro, 1966.
7. Urbain, Peremans and Philippart, 1967.
8. Philippart and Franceschetti, 1967.
9. Whitfield, 1969.
10. Langelaan, 1969.
11. Percy and Brady, 1968.
12. Leroy, Dumon and Radermecker, 1970.
13. Nadler and Gerbie, 1970.
14. Mapes *et al.*, 1970, cited in 15.
15. cited in Nature, 1970.
16. Schneck, Friedland, Valenti, Adachi, Amsterdam and Volk, 1970.
17. Kint, 1970.
18. Wherrett, 1967.

the specific enzyme activity is low in the tissues in some instances it has also been shown to be low in the plasma.

Cultured fibroblasts from skin biopsy might be expected to contain the particular enzyme and hence skin cultures from patients should allow a deficiency to be demonstrated. This has proved to be so in some of the diseases so far studied.

Urine lipid. The presence of metachromatically staining lipid in the cells of the urinary sediment has long been used in the diagnosis of metachromatic leucodystrophy although the value of the test has varied in different hands. It was extended by extracting the lipid from the urinary sediment and subjecting it to paper or thin-layer chromatography. This approach has now been shown to be useful in several of the diseases in the present group.

Cells from amniotic fluid, either as first obtained or after culture, will be increasingly important in the antenatal diagnosis of these conditions and already there are examples where diagnoses have been made *in utero* by examination of this material for the appropriate enzyme activity.

The application of both lipid and enzyme investigations to these various tissues and fluids that have so far been demonstrated in the several diseases are summarised in Table 11.1. Where there are still blank squares in this table the reason is more likely to be because the particular investigation has not yet been made rather than because it has proved to be negative.

Because new diseases undoubtedly remain to be discovered it is important that all cases are examined as fully as possible. The most valuable investigation at present is assay of the enzyme activity in circulating leucocytes but where possible this should be followed by investigation of the isoenzyme pattern. In most instances the abnormal lipid should be sought for and the urinary sediment is proving a useful source of this. It would appear that some conditions characterised by the storage of different lipids and due ultimately to a deficiency in different enzymes are otherwise clinically quite indistinguishable and it is only by continuing to investigate each case fully that any differences in natural history will be revealed.

ACKNOWLEDGEMENTS

Our own investigations in this field, which are associated with Lea Castle Hospital, Wolverley, have been contributed to by Dr. D. E. Langelaan and Mrs. Gillian Haynes. We are indebted to the Endowment Fund of the United Birmingham Hospitals and to the Birmingham Regional Hospital Board for financial support.

REFERENCES

Griffiths, M. I., Langelaan, D. E. and Raine, D. N. (1970). In 'Errors of Phenylalanine, Thyroxine and Testosterone Metabolism' (eds W. Hamilton and F. P. Hudson), p. 36. Edinburgh: E. and S. Livingstone.

Jatzkewitz, H. (1969). In 'Some Inherited Disorders of Brain and Muscle' (eds J. D. Allan and D. N. Raine), p. 114. Edinburgh: E. and S. Livingstone.

Kampine, J. P., Brady, R. O., Kanfer, J. N., Feld, M. and Shapiro, D. (1966). *Science* **155,** 86.

Kint, J. A. (1970). *Science* **167,** 1268.

Langelaan, D. E. (1969). *Biochem. J.* **112,** 26P.

Leroy, J. G., Dumon, J. and Radermecker, J. (1970). *Nature* **226,** 553.

Mapes *et al.* (1970). *FEBS Letters* **7,** 180.

Nadler, H. L. and Gerbie, A. B. (1970). *New Eng. J. Med.* **282,** 596.

Nature, (1970). More clues from enzymes. *Nature* **226,** 596.

O'Brien, J. S. (1969a). *Nature* **224,** 1038.

O'Brien, J. S. (1969b). *Lancet* **2,** 805.

Okada, S. and O'Brien, J. S. (1969). *Science* **165,** 698.

Percy, A. K. and Brady, R. O. (1968). *Science* **161,** 594.

Philippart, M. and Franceschetti, A. T. (1967). *Lancet* **2,** 1368.

Raine, D. N. (1969a). In 'Some Inherited Disorders of Brain and Muscle' (eds J. D. Allan and D. N. Raine), p. 89. Edinburgh: E. and S. Livingstone.

Raine, D. N. (1969b). *Lancet* **2,** 959.

Raine, D. N. (1970). *Devel. Med. Child Neurol.* **12,** 348.

Sandhoff, K., Andreae, U. and Jatzkewitz, H. (1968). *Life Sciences* **7,** 283.

Schneck, L., Friedland, J., Valenti, C., Adachi, M., Amsterdam, D. and Volk, B. W. (1970). *Lancet* **1,** 582.

Suzuki, K. and Suzuki, Y. (1970). *Proc. Nat. Acad. Sci. U.S.* **66,** 302.

Thomas, G. H. (1969). *J. Lab. clin. Med.* **74,** 725.

Urbain, G., Peremans, J. and Philippart, M. (1967). *Lancet* **1,** 1111.

Wherrett, J. R. (1967). *Clin. chim. Acta* **16,** 135.

Whitfield, A. E. (1969). *Biochem. J.* **112,** 26P.

Wolfe, L. S., Callahan, J., Fawcett, J. S., Andermann, F. and Scriver, C. R. (1970). *Neurol.* **20,** 23.

DISCUSSION

Myant: Do you get accumulation of G_{M2} in fibroblasts from Tay-Sachs disease?

Raine: No, but you can demonstrate the enzyme deficiency in fibroblasts.

Robinson: I think that one should be careful about accepting absence of enzyme activity in the urine as a diagnostic test. Activity may be lost on standing or may be depressed by inhibitors.

Scriver: I should like to describe an interesting application of enzyme assay in G_{M1} gangliosidosis which helped to resolve some human psychiatric problems. In a Mohawk Indian family one child had died, and in another we established the diagnosis of G_{M1} generalised gangliosidosis on clinical and chemical grounds and later demonstrated absence of β-galactosidase activity in the leucocytes. A normal brother and sister with excellent school records, began to deteriorate in academic performance and school attendance. We investigated them and found normal leucocyte β-galactosidase activity. After we had reassured them that they did not carry the abnormal gene their psychiatric problems rapidly resolved.

Komrower: I have come across several similar situations. Even where there is no treatment for a disorder we can sometimes help considerably by establishing a definite diagnosis.

Benson: The two slowly migrating bands of β-galactosidase (types B and C) have been reported to be deficient in at least five genetically distinct disorders (Mucopolysaccharidoses types I-III (Ho and O'Brien, 1969, *Science* **165**, 611; generalised gangliosidosis G_{M1}, type 1 (where all bands of β-galactosidase are deficient) and type 2 (deficiency of types B and C only–O'Brien, 1969, *Lancet* **2**, 805) and one is therefore led to consider that the defect either affects different forms of the enzyme, which have distinct substrate specificities or that in at least some of these conditions the deficiency may be secondary.

Using 4-methylumbelliferyl-β-D-galactopyranoside (MUGal) as substrate we have found that cultured fibroblasts from skin have a main band of β-galactosidase activity, on starch gel electrophoresis, corresponding in mobility to that of liver type B and a minor band corresponding to type C. Cultured skin fibroblasts from patients with Hurler and Sanfilippo syndromes, which exhibited metachromasia had at least normal activities of β-galactosidase using MUGal as substrate and normal types B and C on electrophoresis (Benson *et al.*, 1970, *New Eng. J. Med.* **283**, 999).

Raine: In three types of storage disease with visceral involvements the storage substances have a terminal galactose. These are gangliosidoses G_{M1} and G_{M3} and Fabry's disease. Suzuki (1968, *Science* **159**, 147) was concerned as to why G_{M2} gangliosidoses did not have visceral involvement. He was able to extract from G_{M1} kidneys a Keratan-like mucopolysaccharide (MPS), and considered that this may be responsible for the visceral manifestations.

Muir: It is very difficult to characterise Keratan sulphate. Most workers have not found accumulations of any specific GAG in the gangliosidoses.

Raine: It is interesting though that Keratan sulphate is the only MPS which has a terminal galactose.

Robinson: Has anyone investigated a possible relation between these storage diseases and intestinal malabsorption syndromes? In the intestine there are two or three slow moving forms and one fast moving form of β-galactosidase. In cases of lactose intolerance one or two of the slow moving forms are missing. Only one of these forms has been shown to be a lactase.

Scriver: There is no evidence that the patients with gangliosidoses have dissacharide intolerance.

Robinson: There is some interesting work by Gray and Santiago (1969, *J. Clin. Invest.* **48**, 716) who have shown that the specificity of kidney and intestinal glycosidases is related to their molecular size. The highly specific lactase is a large molecule but the smaller enzymes are relatively non specific glycosidases acting on lactose, other galactosides and even xylosides and α-arabinosides. If in an individual one of these enzymes is missing, sometimes some of the other types are also absent and this suggests that they may be related genetically.

Allison: This might be due to the same subunit being present in a number of different enzymes.

Muir: Do patients with gangliosidoses and renal involvement show any features of renal functional impairment?

Raine: No. Renal function appears to be normal, although in our case of globoside storage disease there is polyuria.

Allison: I am worried about the use of synthetic substrates for assay of enzymes normally acting on complicated substrates like gangliosides.

Muir: The natural substrate, G_{M1}, has also been used to demonstrate β-galactosidase deficiency in generalised gangliosidosis in liver, spleen, kidney, brain (Okada and O'Brien, 1968, *Science* **160**, 1002) and in fibroblasts (Sloan *et al.*, 1969, *Pediat. Res.* **3**, 532).

Raine: Yes, I agree but natural substrates are normally not available. An interesting consideration for leucocyte enzyme studies is the method of expression of enzyme units.

Percy and Brady (1968, *Science* **161**, 594), who described leucocyte arylsulphatase deficiency at about the same time as ourselves, expressed their results according to units of activity per mgm of leucocyte protein and were unable to show intermediate levels of activity in heterozygotes which we did by relating leucocyte arylsulphatase activity to the activity of another leucocyte enzyme.

Session IV

Chromosomal Redundancy associated with Mental Subnormality

Biochemistry of Down's Syndrome

J. STERN

Queen Mary's Hospital for Children, Carshalton, Surrey

The clinical manifestations of Down's syndrome have been recognised and studied for over a century and are well documented (Penrose and Smith, 1966). The disease was for a long time erroneously regarded as a variant of cretinism, and a biochemical or endocrine basis of the disease has been suspected ever since. Disappointingly, in spite of much research, no biochemical abnormality has yet been found which is both pronounced and present in every case. However, patients with Down's syndrome as a group do differ significantly in many aspects of their biochemistry from normal subjects and other mentally retarded patients. Following the discovery of the extra chromosome in the disease more than a decade ago, interest has been focused on the question if, and to what extent, any biochemical abnormalities are a direct consequence of the effect of the chromosomal aberration on the metabolism of cells and the organism. The most challenging and at the present largely unsolved problem is that of establishing if any of the biochemical abnormalities can be related to the mental retardation which, after all, is still much the most important handicap of the patient with Down's syndrome.

Anyone following the literature on the biochemistry of Down's syndrome cannot fail to notice the number of reported abnormalities which were not confirmed in subsequent investigations. Many difficulties undoubtedly arise from the cross-sectional design of most studies. In these one or more biochemical parameters are measured in a group of patients with Down's syndrome and compared to normal or mentally retarded controls. If this approach is used differences are almost certain to be discovered; the only snag is that many of these are apt to turn out to be artefacts. Even if patients are matched for age, sex, weight and mental age other variables remain which are more difficult to control such as dietary preferences, mobility, undiagnosed infections and additional handicaps such as congenital disease of the heart in patients with Down's syndrome. Malabsorption may occur as the aftermath of a gastrointestinal infection. Artefacts are by no means confined to institutionalised patients. Biochemical findings in patients living at home may reflect deprivation due to poverty or faulty rearing, such as prolonged and unsupplemented bottle feeding.

The levels of some metabolites may fluctuate during the period immediately following admission to a ward, which for many retarded patients is a time of acute stress.

Differences between patients with Down's syndrome and controls may be limited to certain age groups, due occasionally to selective mortality of the more severely affected patients or, more often, to unequal rates of maturation and ageing. Examples are the serum cholesterol level which is higher than normal in children with Down's syndrome aged 2-6 years, but not in those aged 6-10 years (Stern and Lewis, 1957a) and the leucocyte alkaline phosphatase which Alter *et al.* (1963) found to be raised in Down's syndrome in children only if they were below the age of 10 years (cf. also Alter and Lee, 1968 and O'Kell, 1968). Occasionally the discrepancies in the results obtained by workers in this field are too great to be explained by the factors mentioned so far, or even by differences in technique. Thus Ong *et al.* (1967) reported a fivefold increase in the serum creatine kinase level of patients with translocation and trisomic Down's disease compared to controls. However, when we tried to repeat this work we could detect no difference whatsoever between a group of patients with Down's syndrome and controls (Chapman and Stern, unpublished) (Fig. 1). It appears that this biochemical parameter is susceptible to unknown, presumably environmental, factors. It follows in general that because of the presence of such unrecognised and therefore uncontrolled factors, biochemical abnormalities should only be accepted as characteristic of the disease when confirmed in several centres, preferably in more than one country. Futhermore, biochemical changes which are claimed to be a direct consequence of the chromosomal redundancy are unconvincing unless shown to be present over a wide age-range.

Serum Proteins

An increase in γ-globulin and decrease in albumin were amongst the earliest biochemical findings in Down's disease (Donner, 1954; Stern and Lewis, 1957b). These observations have in general beem confirmed by later workers. Minor differences in results can probably be attributed to one or other of the factors outlined in the previous section.

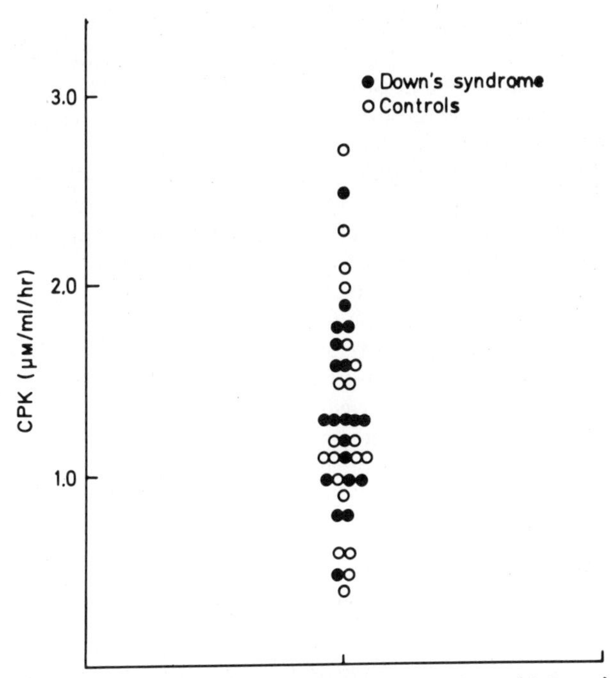

Fig. 1. Serum creatine phosphokinase levels in children with Down's syndrome and mentally retarded controls.

It seemed from these studies that patients with Down's syndrome respond to antigenic and probably other stimuli with an excessive and persistent rise in γ-globulin. On the other hand, patients with Down's syndrome have a high rate of infection which suggests the possibility of a deficiency of one or other group of immunoglobin. Adinolfi *et al.* (1967) established that patients with Down's syndrome as a group have raised IgG levels; the IgM level on the other hand was significantly lower. IgA levels were normal. No evidence was found for the production of 'faulty' immunoglobulins as had been postulated by some authors. Since bacterial antibodies are associated with IgG, IgA and IgM globulins, chronic infection would have increased all groups of immunoglobulins and would not therefore be invoked to explain the observed immunoglobulin pattern which was more in keeping with that observed in patients with disorders involving the lymphocytes or reticulo-endothelial system. Miller *et al.* (1969) who studied immunoglobulins in newborn babies with Down's syndrome found that the IgG level was depressed compared to controls. Little IgA was found in either group and IgM levels were similar. As the mothers' IgG levels were normal, placental transport of IgG appeared impaired. Bearing in mind the immunoglobulin pattern in older patients, Miller *et al.* suggest that in Down's syndrome the increased IgG level might be compensatory to a primary defect in IgM synthesis.

Agarwal *et al.* (1970) studied DNA polymerase activity as an index of lymphocyte stimulation in Down's syndrome and found a comparatively poor response after PHA stimulation. This may reflect impairment of cellular immune function. Deficiencies in both humoral and cellular immune response may therefore contribute to the increased susceptibility to infection in Down's syndrome.

An increased incidence of raised IgM levels in the cord blood of newborn babies with trisomies has been reported by Robinson *et al.* (1969). These authors inferred that the high IgM levels are the reflection of intra-uterine infection, and suggested that this fitted their theory that viral infections around the time of conception might be one of the factors producing non-disjunction.

Tryptophan and Vitamin B_6 Metabolism

Abnormalities at various levels of the nicotinamide pathway of tryptophan metabolism have been reported by many workers and reviewed by Penrose and Smith (1966) and Crome and Stern (1967). Children with Down's syndrome were in most studies found to excrete less xanthurenic acid after an oral tryptophan load than mentally retarded control subjects. When given the pyridoxine antagonist deoxypyridoxine patients with Down's syndrome responded with an increased excretion of xanthurenic acid and other metabolites of the nicotinamide pathway, and of oxalic acid, compared with matched controls (McCoy *et al.,* 1964). McCoy *et al.* concluded that this indicated a tendency to vitamin B_6 depletion. The fact that studies of the nicotinamide pathway have often yielded contradictory results is not surprising. The tryptophan load test is notoriously beset by methodological difficulties (Symposium, 1969). These include substrate induction of tryptophan pyrrolase, the first enzyme of the pathway, and of other pyridoxine dependent enzymes. The urinary excretion pattern also depends on the dietary intake of tryptophan prior to the load, the size of the test dose, and is affected by hormones. The decreased taurine excretion in Down's syndrome (Wainer *et al.,* 1966; King *et al.,* 1968) was stated to correlate with adaptive behaviour. It is now attributed to increased renal tubular reabsorption rather than B_6 deficiency (McCoy *et al.,* 1969). Direct assessment of the vitamin B_6 state of the patient is difficult because of the occurrence of a number of B_6 vitamers sometimes bound to protein (Symposium, 1969). Studies of the urinary excretion of 4-pyridoxic acid, a major metabolite of vitamin B_6, again suggest that patients with Down's syndrome have low stores of the vitamin. The level of pyridoxal phosphate in the leucocytes of patients with Down's syndrome was found to be low (Coburn and Seidenberg, 1969) but this result may not reflect the situation in the tissues in general as some of the metabolite was probably hydrolysed by the active leucocyte alkaline phosphatase.

Abnormalities undoubtedly also exist in the serotonin pathway of tryptophan metabolism and these have attracted much interest because of the role of serotonin in brain metabolism. The blood levels in trisomic Down's disease are greatly reduced (Bazelon, 1967). Translocation patients were said to have intermediate levels, but this is not generally accepted. Serotonin is formed from 5-hydroxytryptophan by a pyridoxine dependent decarboxylase. Deficiency in apoenzyme, cofactor or substrate could result in low serotonin levels. In my view the susceptibility to vitamin B_6 antagonist, the low blood serotonin level, and possibly other abnormalities are caused by alteration in the absorption or transport of tryptophan and its metabolites and perhaps also of vitamin B_6 in the intestine, at the cell membrane or in the kidney (cf. also Bazelon, 1967). The low blood serotonin is reflected in the platelet level (McCoy *et al.,* 1968). There is also some evidence that the binding of serotonin to platelets is defective (Boullin *et al.,* 1969). Defective serotonin binding would have important implications if it also occurred in brain. The end product of the serotonin pathway, 5-hydroxyindole acetic acid, is normal in cerebrospinal fluid (Dubowitz and Rogers, 1969); there are no data as yet on other metabolites of the serotonin pathway in CSF. The practical importance of serotonin in Down's syndrome is that administration of its precursor, 5-hydroxytryptophan, reversed the hypotonia in infants with the disease (Bazelon *et al.,* 1967). Intelligence was found not to be correlated to the blood serotonin level by Marsh (1969) in five cases of Down's disease.

Enzymes of the Formed Elements of the Blood

Many studies have been undertaken in an effort to demonstrate a gene dosage effect due to the extra chromosome 21 in Down's disease. The first suggestion made was that the gene for leucocyte alkaline phosphatase might be located on chromosome 21. Subsequently more than two dozen white and red cell enzymes have been investigated. While deviations from normal have been found in more than half the enzymes, no simple gene dosage effect has been unequivocally demonstrated. White cell enzymes studied in addition to alkaline phosphatase include acid phosphatase, galactose 1-phosphate uridyl transferase and glucose 6-phosphate dehydrogenase. Increased levels compared to controls have been found for all these enzymes, and the difference in activity persisted when lymphocytes were stimulated with phytohaemagglutinin (Nadler *et al.*, 1967a, b). There is, however, considerable overlap in the values for the groups quite apart from intra-individual variability (Phillips *et al.*, 1967). Nor is the increase in enzyme activity found in platelets (Shih and Hsia, 1966) or cultured fibroblasts (Nadler *et al.*, 1967c; cf. also Hsia *et al.*, 1968). As some of the white cell enzymes are adaptive, some workers have preferred to study the enzymes of the red cells, as in the absence of a cell nucleus interpretation of the results might be expected to be easier. Raised enzyme levels undoubtedly occur for a number of enzymes, but again results have been variable. At least one of the enzymes, glucose 6-phosphate dehydrogenase, is known to be located on the X chromosome, and its level depends on the age of the erythrocyte, as does that of a number of other enzymes. Table 12.1 shows assays of red cell enzymes in children aged 4-10 years and in newborns by Pantelakis *et al.* (1970). While four enzymes were raised in the older children only two enzymes were increased in the newborn with Down's disease. Of particular interest is the increase of approximately 50% in phosphofructokinase which appears to be the only red cell enzyme whose activity is consistently raised. Pantelakis *et al.*, favour a gene dosage effect as an explanation. In support of this view is that a smaller increase in the activity of the enzyme is found in mosaic Down's syndrome (Benson *et al.*, 1968). However, phosphofructokinase is a key enzyme in the control of carbohydrate metabolism and as such susceptible to several control mechanisms (Lowry and Passoneau, 1966). This enzyme is, for example, activated by an increase in inorganic phosphate, fructose 6-phosphate, AMP, ADP and NH_4^+ and inhibited by ATP and citrate. A moderate change in the proportions of these activator and inhibitor molecules could result in a pronounced change in enzymic activity. Platelet phosphofructokinase is normal in Down's disease (Doery *et al.*, 1968).

Sparkes and Baughan (1969) studied blood cell enzymes in D/21 and G/21 translocation patients, on the supposition that in translocation Down's syndrome most genes would be present in triplicate but that some might be lost in the formation of the translocation, resulting in partial monosomy on the D,21 or 22 chromosome. The only evidence for such a mechanism was found in two patients with D/21 translocation who had a reduced level of phosphofructokinase. In general, however, Sparkes and Baughan concluded that enzyme studies

TABLE 12.1

Increase in the activity of Red Cell enzymes in patients with Down's Syndrome compared to controls (adapted from Pantelakis *et al.*, 1970)

Enzyme	*Newborn Infants*		*Children aged 4-12 years*	
	Increase in activity (per cent of normal)	*Significance*	*Increase in activity (per cent of normal)*	*Significance*
Hexokinase	1		48	$t = 3.5, P < 0.005$
Phosphohexokinase	75	$t = 5.9, P < 0.001$	70	$t = 7.2, P < 0.001$
Pyruvate kinase	–12		15	
Lactic dehydrogenase	2		6	
Glucose-6-phosphate dehydrogenase	–13		–18	
6-phosphogluconate dehydrogenase	62	$t = 3.2, P < 0.005$	93	$t = 7.0, P < 0.001$
Catalase	4		7	
NADPH methaemoglobin reductase	4		33	$t = 3.0, P < 0.005$

did not help in localising the genes. The control of enzyme activity in higher organisms is complex (Boyer, 1970). In addition to structural and regulator genes the enzyme level is affected by activators and inhibitors, substrate induction and feedback inhibition, catabolism and excretion of the enzyme, and hormonal control mechanisms. In studies with peripheral leucocytes additional difficulties arise. Preparation of white cells free from red cells is difficult, and, furthermore, marked differences may exist between the enzymic activities of lymphocytes and polymorphonuclear leucocytes. The activity of some enzymes depends on the age of the blood cell or patient. Elsewhere in this symposium Professor Yielding points out that, depending on the molecular model chosen, an increase, decrease or no change may result in the activity of an enzyme whose gene is located on a trisomic chromosome.

In leucocytes at least, increased enzyme activity seems to reflect a more generalised disturbance of metabolism. Evidence for such a disturbance are ultrastructural abnormalities (Smith, Penrose and O'Hara, 1967), the increased turnover of cholesterol in lymphocytes (McCoy and Nance, 1969), the greater increase in leucocyte alkaline phosphatase activity after prednisolone (McCoy and Ebadi, 1966), the poor response of DNA polymerase of lymphocytes after PHA stimulation already referred to (Agarwal *et al.*, 1970) and the increased rate of protein and RNA synthesis found by Benson (1967). Further observations by Dr. Benson on RNA synthesis are reported elsewhere in this symposium.

While Galbraith and Valberg (1966) found no difference in granulocyte turnover in Down's syndrome, Mellman *et al.* (1967) subsequently published data indicating clearly a shortened granulocyte half-life in the disease. These authors also produced evidence that the half-life of the granulocyte is inversely related to the uric acid excretion when corrected for body weight.

Uric acid excretion and the serum uric acid level have attracted attention for their inherent interest. Earlier reports suggested extremely high values; in subsequent studies much lower values were found, but patients with Down's syndrome still had significantly higher serum uric acid levels than other mentally retarded patients. Still smaller increases have been reported for the precursors of uric acid, xanthine and hypoxanthine (Appleton *et al.*, 1969). Decreased renal efficiency in handling urate has been inferred from these data. Appleton *et al.*, postulate that vitamin A deficiency interferes with the keratinisation of the renal epithelial cells. In Down's disease urinary uric acid excretion, therefore, appears to be increased by the raised turnover of white cells, and decreased by a relative renal insufficiency.

I believe that many of the biochemical abnormalities in Down's syndrome can be interpreted as due to defective absorption, transport or excretion of metabolites. Earlier investigations of Down's syndrome have mostly been on blood and urine, but more recently interest has extended to other body fluids. An abnormal protein pattern in tears characterised by an increase in the α-glubulins, particularly the α_1-globulin, has been attributed by Allerhand *et al.* (1963) to an abnormality in protein transport. Lindsten (1967), however, points out that this pattern is not confined to patients with Down's syndrome; it also occurs in normal subjects with infections of the respiratory tract.

Baar and Gordon (1963) drew attention to a defect in cation fluxes in red cells. The sodium influx and potassium efflux were significantly higher than in controls. This may again be an age effect as cation transport in red cells is strongly age-dependent. Another defect in cation transport has since been found. Patients with Down's syndrome have little or no dental caries, although they may have severe periodontal disease. This prompted Winer *et al.* (1965) to examine the composition of the parotid saliva of these patients. After citric acid stimulation they found a significant increase in pH, sodium, calcium and bicarbonate compared to other mentally retarded children, but with considerable overlap of values. We thought that differences in sodium reabsorption might be more pronounced at lower flow rates and reinvestigated the sodium level and pH of the unstimulated secretion of the parotid using suitably sensitive glass electrodes (Chapman *et al.*, 1967). The results are illustrated in Fig. 2 together with data for patients with cystic fibrosis of the pancreas and mentally retarded control subjects. The overlap is much smaller than with stimulated secretion. Raised sodium levels in parotid saliva are thus common to patients with Down's disease and cystic fibrosis of the pancreas. The calcium level and pH also appear to be raised in both conditions, in contrast to the slightly reduced calcium level in serum commonly found in Down's disease. Abnormalities in pupillary reactivity are also common to both disorders. However, while the sweat sodium is very high in cystic fibrosis, we have always found it to be normal in Down's disease. No satisfactory explanation for this selective defect of sodium transport in the parotid gland can be offered at present. Wiesmann *et al.* (1970) suggest that saliva from small salivary glands on the oral mucosa is largely responsible for the raised levels of sodium in parotid saliva from patients with cystic fibrosis. If substantiated these considerations would be relevant also in the case of Down's syndrome. In this context it is of interest that the suggestion is being made (Bartman *et al.*, 1970) that tissues in cystic fibrosis are affected by a defect which impairs cellular transport mechanisms and membrane permeability. Cystic

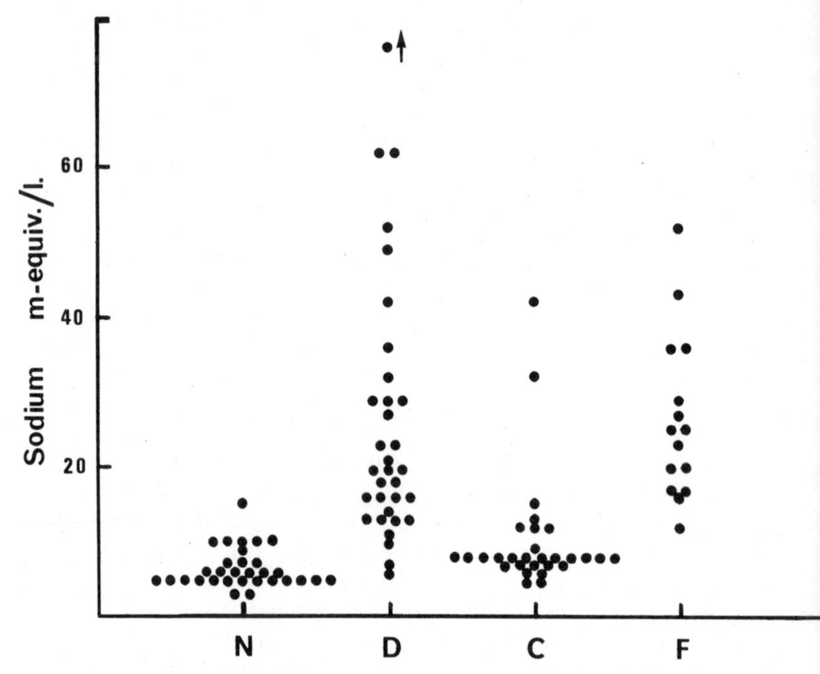

Fig. 2 Sodium levels of unstimulated parotid saliva: N, normal adults; D, children with Down's syndrome; C, severely subnormal children; F, children with cystic fibrosis of the pancreas (with permission from Chapman *et al.*, 1967).

fibrosis and Down's disease in the same patient have been described in at least four cases (Milunsky, 1968).

Workers in Czechoslovakia have recently reported a high incidence of protein losing enteropathy in patients with Down's syndrome (Saxl and Hrstka, 1968), the main losses being in the immunoglobulin IgA, with a smaller loss in IgG. Faecal fat excretion was normal in most cases, xylose absorption abnormal in about one third of their patients. Electron microscopy of the small intestine revealed abnormalities of the brush border, paucity of lysosomes in the cytoplasm and degenerative changes in some mitochondria (Saxl *et al.*, 1968).

It is necessary to point out that some of the absorption defects in Down's diseases are not specific. Thus we found low xylose absorption in a third of patients with Down's syndrome, in agreement with the Czechoslovak authors. A similar proportion of tests was, however, also abnormal in mentally retarded control subjects (Chapman *et al.*, 1966). Flat oral glucose tolerance curves are common not only in patients with

Down's syndrome but also in other mentally retarded patients. Other load tests are often abnormal in Down's disease. Examples are quoted by Crome and Stern (1967).

Endocrinological Aspects of Down's Syndrome

A hundred years ago, because of its superficial similarity to cretinism, Down's syndrome was regarded as a variant of that disease. The widely held suspicion of some endocrine pathology in Down's syndrome is reflected in numerous abortive attempts to treat patients with hormones. With the advent of more sophisticated techniques the scope of hormone assays has greatly increased. Milunsky *et al.* (1968) found no difference in the plasma insulin response to a glucose load between children with Down's syndrome, mentally retarded and normal children. Growth hormone levels determined by radioimmunoassay were normal. Using the same technique Hillman (1969) found normal thyroid-stimulating hormone levels in the disease. Thyroxine measured by competitive protein binding analysis was also normal. Resting levels of some hormones are low, but there is usually good response in stimulation tests (Reiss *et al.,* 1965). This is generally true of mentally retarded children (Hillman *et al.,* 1968). Urinary levels of testosterone and epitestosterone in adult males with Down's syndrome while lower than in male controls nevertheless indicate considerable testicular function (Peters and Culley, 1969). The incidence of diabetes is significantly increased in Down's disease (Milunsky and Neurath, 1968). Some abnormalities in glucose tolerance in the parents have also been reported (Milunsky, 1970).

An increased frequency of thyroid antibodies has been reported both in patients with Down's disease and their mothers (Fialkow and Uchida, 1968; Fialkow, 1969). The antibodies or related factors might themselves be predisposing agents to non-disjunction, or a virus might lead both to maternal antibodies and the chromosomal aberration in the child. Diabetics also have a greater than normal tendency to autoantibody formation, so that if autoimmunity is implicated in non-disjunction an association of diabetes and Down's syndrome is not unexpected (Milunsky, 1970).

The Brain in Down's Syndrome

All hitherto detected biochemical abnormalities in Down's syndrome are, compared to those in the inborn errors of metabolism, quantitative rather than qualitative, and mostly non-specific. Virtually nothing is known about possible biochemical causes of the pathological changes in the brain. These are diffuse and rather variable but are sufficiently severe to account fully for the mental retardation which is invariably present in patients with the disease (Crome and Stern, 1967). Brain weight is somewhat reduced. The frontal lobes, brainstem and cerebellum are particularly small. The middle lobes of the cerebellum and some major tracts of the brainstem are functionally and anatomically related to the frontal lobes. The size of these formations is therefore interdependent. All are relatively small in the newborn and grow more rapidly than other parts of the brain in early infancy. Neurohistological changes are uncertain: both sparseness of cells and excessive neuronal density have been reported. Lesions characteristic of ageing–senile plaques and the Alzheimer neurofibrillary change occur in Down's disease at a comparatively early age. Disorders of migration are frequently present, particularly tuberosities of the medial part of the peduncle of the cerebellar flocculus and rarely, in severely affected patients, microgyria and pachygyria. None of these changes is specific.

Chemical analysis of the cerebral lipids by Stephens and Menkes (1969) revealed no major abnormality. These authors consider that the chromosomal defect results in a metabolic derangement throughout post-conceptual life and in constant interference with the regulation of protein synthesis. The cerebral disorder would then result from faulty pre- and post-natal tissue differentiation and not be reflected in the structure of the lipids. The absence of gross hormonal disturbance in the older patient with Down's disease does not exclude the possibility of hormonal disequilibrium at an earlier, critical stage of development of the brain. The reduction in the size of the cerebellum and related formations appears to be compatible with interference extending to a comparatively late stage of the development of the brain. In the rat, for example, a similar reduction in brain weight, also accentuated in the cerebellum, can be produced when the developing brain is exposed to thyroid deficiency, thyroid or corticosterone excess, or undernutrition. Thyroid deficiency also affects cell division and migration, and protein and enzyme synthesis (Balázs, 1970).

Abnormalities in transport of hormones synthesised extracerebrally or of essential amino acids could affect the induction of cerebral enzymes at critical stages of development and interfere with differentiation and cell migration. Hormonal action is sometimes mediated by cyclic AMP and the level of this compound is increased by serotonin (Kakiuchi *et al.*, 1970). This is one mechanism by which an abnormality of serotonin metabolism in Down's syndrome could distort the biochemical development of the brain. McIlwain (1970, 1971) has described a number of enzymatically mediated adaptations of which the brain is capable and which are indeed essential for its normal development. Interference with metabolic adaptation may be an important way by which the extra chromosome affects the brain during intrauterine development.

REFERENCES

Adinolfi, M., Gardner, B. and Martin, W. (1967). *J. clin. Path.* **20,** 860.

Agarwal, S. S., Blumberg, B. S., Gerstley, B. J. S., London, W. T., Sutnick, A. I. and Loeb, L. A. (1970). *J. clin. Invest.* **49,** 161.

Allerhand, J., Karelitz, S., Isenberg, H. D., Penbharkul, S. and Ramos, A. (1963). *J. Pediat.* **62,** 235.

Alter, A. A. and Lee, S. L. (1968). *Ann. N.Y. Acad. Sci.* **155,** 1023.

Alter, A. A., Lee, S. L., Poulfar, M. and Dobkin, G. (1963). *Blood* **22,** 165.

Appleton, M. D., Haab, W., Burti, U. and Orsulak, P. J. (1969). *Amer. J. ment. Defic.* **74,** 196.

Baar, H. S. and Gordon, M. (1963). Proceedings of the 2nd International Congress on Mental Retardation, Vienna 1961, Part 1. Basel: Karger, p. 373.

Balázs, R. (1970). In 'Cellular Aspects of Growth and Differentiation in the Developing Brain' (ed. D. C. Pease). UCLA Forum of the Medical Sciences, Los Angeles: University of California Press.

Bartman, J., Wiesmann, U. and Blanc, W. A. (1970). *Pediatrics* **76,** 430.

Bazelon, M. (1967). *Clin. Proc. Child. Hosp. Wash.* **23,** 58.

Bazelon, M., Paine, R. S., Cowie, V. A., Hunt, P., Houck, J. C. and Mahanand, D. (1967). *Lancet* **i,** 1130.

Benson, P. F. (1967). *Nature* **215,** 1290.

Benson, P. F., Linacre, B. and Taylor, A. I. (1968). *Nature* **220,** 1235.

Boullin, D. J., Coleman, M. and O'Brien, R. A. (1969). *J. Physiol.* **204,** 128P.

Boyer, S. H. (1970). In 'Modern Trends in Human Genetics' (ed. A. E. H. Emery), vol. 1. London: Butterworth, p. 1.

Chapman, M. J., Donoghue, E. C., Saggers, B. A. and Stern, J. (1967). *J. ment. Defic. Res.* **11,** 185.

Chapman, M. J., Harrison, P. M. and Stern, J. (1966). *J. ment. Defic. Res.* **10,** 19.

Coburn, S. P. and Seidenberg, M. (1969). *Amer. J. clin. Nutr.* **22,** 1197.

Crome, L. and Stern, J. (1967). 'The Pathology of Mental Retardation'. London: Churchill.

Doery, J. C. G., Hirsch, J., Garson, O. M. and De Gruchy, G. C. (1968). *Lancet* **ii,** 894.

Donner, M. (1954). *Ann. Med. exp. Fenn.* **32,** supp. 9.

Dubowitz, V. and Rogers, K. J. (1969). *Develop. Med. Child. Neurol.* **11,** 730.

Fialkow, P. J. (1969). In 'Progress in Medical Genetics' (eds A. G. Steinberg and A. G. Bearn), Vol. 9, p. 117. New York: Grune and Stratton.

Fialkow, P. J. and Uchida, I. A. (1968). *Ann. N.Y. Acad. Sci.* **155,** 759.

Galbraith, P. R. and Valberg, L. S. (1966). *Pediatrics* **37,** 108.

Hillman, J. C. (1969). *J. ment. Defic. Res.* **13,** 191.

Hillman, J., Hammond, J., Sokola, J. and Reiss, M. (1968). *J. ment. Defic. Res.* **12,** 294.

Hsia, D. Y-Y., Nadler, H. L. and Shih, L-Y. (1968). *Ann. N.Y. Acad. Sci.* **155,** 716.

Kakiuchi, S., Rall, T. W. and McIlwain, H. (1970). *J. Neurochem.* **16,** 485.

King, J. S., Goodman, H. O., Wainer, A. and Thomas, J. J. (1968). *J. Nutr.* **94,** 481.

Lindsten, J. (1967). In 'Mongolism' (eds G. E. W. Wolstenholme and R. Porter). Ciba Foundation Study Group No. 25. London: Churchill, p. 86.
Lowry, O. H. and Passoneau, J. V. (1966). *J. biol. Chem.* **241,** 2268.
McCoy, E. E., Anast, C. S. and Naylor, J. J. (1964). *J. Pediat.* **65,** 208.
McCoy, E. E., Colombini, C. and Ebadi, M. (1969). *Ann. N.Y. Acad. Sci.* **166,** 116.
McCoy, E. E. and Ebadi, M. (1966). *J. Pediat.* **68,** 835.
McCoy, E. E. and Nance, J. L. (1969). *J. ment. Defic. Res.* **13,** 34.
McCoy, E. E., Rostafinsky, M. J. and Fishburn, C. (1968). *J. ment. Defic. Res.* **12,** 18.
McIlwain, H. (1970a). *Nature* **226,** 803.
McIlwain, H. (1971). This symposium p. 15 *et seq.*
Marsh, R. W. (1969). *N.Z. med. J.,* **70,** 179.
Mellman, W. J., Raab, S. O. and Oski, F. A. (1967). In 'Mongolism' (eds G. E. W. Wolstenholme and R. Porter). Ciba Foundation Study Group No. 25. London: Churchill, p. 77.
Miller, M. E., Mellman, W. J., Cohen, M. M., Kohn, G. and Dietz, W. H. (1969). *J. Pediat.* **75,** 996.
Milunsky, A. (1968). *Pediatrics* **42,** 501.
Milunsky, A. (1970). *Amer. J. ment. Defic.* **74,** 475.
Milunsky, A., Lowy, C., Rubenstein, A. H. and Wright, A. D. (1968). *Develop. Med. Child. Neurol.* **10,** 25.
Milunsky, A. and Neurath, P. W. (1968). *Arch. environm. Hlth.* **17,** 372.
Nadler, H. L., Monteleone, P. L., Inouye, T. and Hsia, D. Y-Y. (1967a). *Blood* **30,** 669.
Nadler, H. L., Monteleone, P. L. and Hsia, D. Y-Y. (1967b). *Life Sci.* **6,** 2003.
Nadler, H. L., Inouye, T. and Hsia, D. Y-Y. (1967c). *Amer. J. hum. Gent.* **19,** 94.
O'Kell, R. R. (1968). *Ann. N.Y. Acad. Sci.* **155,** 980.
Ong, B. H., Rosner, F., Mahanand, D., Houck, J. C. and Paine, R. S. (1967). *Develop. Med. Child Neurol.* **9,** 307.
Pantelakis, S. N., Karaklis, A. G., Alexiou, D., Vardas, E. and Valaes, T. (1970). *Amer. J. hum. Genet.* **22,** 184.
Penrose, L. S. and Smith, G. F. (1966). 'Down's Anomaly.' London: Churchill.
Peters, A. and Culley, W. (1969). *Clin. chim. Acta* **25,** 199.
Phillips, J., Herring, R. M., Goodman, H. O. and King, J. S. (1967). *J. med. Genet.* **4,** 268.
Reiss, M., Wakoh, T., Hillman, J. C., Pearce, J. J., Daley, N. and Reiss, J. M. (1965). *Amer. J. ment. Defic.* **70,** 204.
Robinson, A., Goad, W. B., Puck, T. T. and Harris, J. S. (1969). *Amer. J. hum. Genet.* **21,** 466.
Saxl, O. and Hrstka, V. (1968). *Mschr. Kinderheilk.* **116,** 230.
Saxl, O., Tichy, J., Hradsky, M. and Hrstka, V. (1968). *Z. Gastroent.* **6,** 11.
Shih, L-Y. and Hsia, D. Y-Y. (1966). *Lancet* **i,** 155.
Smith, G. F., Penrose, L. S. and O'Hara, P. T. (1967). *Lancet* **ii,** 452.
Sparkes, R. and Baughan, M. A. (1969). *Amer. J. hum. Genet.* **21,** 430.
Stephens, M. C. and Menkes, J. H. (1969). *Develop. Med. Child Neurol.* **11,** 346.
Stern, J. and Lewis, W. H. P. (1957a). *J. ment. Defic. Res.* **1,** 96.
Stern, J. and Lewis, W. H. P. (1957b). *J. ment. Sci.* **103,** 222.
Symposium (1969). 'Vitamin B_6 in Metabolism of the Nervous System' (ed. M. A. Kelsall). *Ann. N.Y. Acad. Sci.* **vol. 166,** Art. 1.
Wainer, A., King, J. S., Goodman, H. O. and Thomas, J. J. (1966). *Proc. Soc. exp. Biol. Med.* **121,** 212.

Wiesmann, U. N., Boat, T. F. and Di Sant'Agnese, P. A. (1970). *J. Pediat.* **76,** 442.
Winer, R. A., Cohen, M. M., Feller, R. P. and Chauncey, H. H. (1965). *J. dent. Res.* **44,** 632.
Yielding, K. L. (1970). This symposium.

DISCUSSION

Scriver: Have there been any biochemical studies on cultured tissues, for example has anyone looked at the ageing of cells from Down's syndrome?
Stern: I am not aware of any work on cell ageing. Hshia and Nadler have shown that unlike in WBC's and RBC's there is normal enzyme activity in cultured fibroblasts from children with Down's syndrome (Nadler *et al.*, 1967, *Am. J. hum. Genet.* **19,** 94).
Polani: There is no clear cut evidence that cultured fibroblasts derived from subjects with Down's syndrome show any differences from controls with respect to ageing. Methods for measuring cell ageing in culture are not very satisfactory however.
Benson: There have been some studies on cultured Down's syndrome fibroblasts. Cox, 1965 (*Exp Cell Res.* **37,** 690) found that activities of fibroblast alkaline phosphates both in the induced and non-induced states were similar in cultures from controls and from Down's syndrome. Mittwoch, 1967 (CIBA Foundation Study Group No. 25 eds Wolstenholme and Porter, 51-60) measured DNA content of individual nuclei and found that there was a higher proportion of cells with DNA values intermediate between diploid and tetraploid in cultures of G-trisomic than in diploid cells. She suggested that G-trisomic cells might require a longer period for DNA synthesis. Kabach and Bernstein, 1969 (paper delivered to N.Y. Academy of Science) observed a slower rate of DNA synthesis during exponential growth only and a faster rate of RNA synthesis by G-trisomic than normal diploid cultured fibroblasts. They suggested that these observations could indicate that the cell cycle of G-trisomic cells was prolonged and that a derangement of RNA production could reflect a 'cellular ageing phenomenon'.

Some years ago (Benson, 1966, a paper delivered to the Paediatric Research Society) I reported a higher rate of incorporation of ^{3}H-uridine into RNA of cultured skin fibroblasts (in the exponential growth phase) derived from subjects with Down's syndrome than from controls. Sucrose density gradient studies revealed that after short incubation periods (0.5-1 hour) most of the labelled RNA was polydisperse but that after longer incubation periods (2-24 hours) it was predominantly ribosomal.

McIlwain: I was interested in the comments by Dr. Stern that ion exchanges and 'sodium pump' mechanisms might be abnormal in Down's syndrome. In studying such phenomena one should be careful not to confuse the mechanisms for the ion pump with those for leakage of ions through membranes by increased permeability. In several systems passive cation movements can be blocked effectively with tetrodotoxin. I should be interested to know if subjects with Down's syndrome have been found to be abnormally sensitive to such toxins.

Milne: I was pleased that Dr. Stern stressed that one should exercise caution in the interpretation of biochemical differences found in Down's syndrome. Certainly in the case of uric acid clearance studies, with which I have had some experience, results can be very easily affected by a number of factors, e.g. hormones. Any differences obtained in other conditions should be accepted as valid only after the results have been very carefully scrutinised.

Polani: Does Professor Rees know of any studies on the effect of chromosomal gain in plants on K^+ or Na^+ metabolism?

Rees: No. I have no information on this subject. I wonder whether the abnormalities in human trisomics have been considered in relation to lack of stability? In plants chromosome gain leads to a wider range of variation in regulatory processes generally.

Polani: Possibly the reverse might be the case. The variability of a number of quantitative parameters such as height and birth weight in subjects with Down's syndrome tends to be significantly less than in normal subjects; but I don't know if this applies to biochemical characteristics.

Allison: I was impressed with the wide range of biochemical values shown by a number of the studies. I think that general factors such as duration of cell survival and possibly influence on enzyme activity by environmental factors such as raised uric acid levels may produce a whole range of abnormal biochemical values. This sort of secondary effect almost certainly occurs in leukaemic cells.

Benson: The degree of variability in values of a biochemical measurement which has been found to be abnormal in Down's syndrome, depends, to a certain extent, on the direction and magnitude of the altered distribution. Thus when the mean of a group of observations is significantly higher in G-trisomics than in controls, the standard deviation also tends to be higher. For example, the rate of incorporation of isotopic precursors into leucocyte RNA and protein (Benson, 1967, *Nature* **215**, 1290). On the other hand when the mean in Down's syndrome is lower than in controls the standard deviation also tends to be lower. For example, platelet monoamine oxidase activity (Benson and Southgate, 1971, *Am. J. hum. Genet.* **23**, 211.

Scriver: Have any longitudinal studies been carried out on biochemical levels in Down's syndrome?

Polani: I don't know of any longitudinal biochemical studies. Mrs. Dicks-Mireaux, Dr. Pampiglione and myself in one study (Dicks-Mireaux, 1970, *Amer. J. ment. Defic.* in the Press; Dicks-Mireaux, Pampiglione and Polani, unpublished) and Dr. Cowie (Cowie, 1970, A study of the early development of mongols, Monograph No. 1 of the Institute for Research into Mental Retardation. Pergamon Press, London and New York) in another followed up newborns with Down's syndrome with respect to development and EFG changes (see also Penrose and Smith, 1966, 'Down's Anomaly.' J. and A. Churchill, London).

Spector: Is the intellectual performance of infants with Down's syndrome impaired at birth?

Polani: Yes, there was definite intellectual impairment at birth. The follow-up also suggested that further intellectual loss occurred during the first two-three years of life. After this there seems to be no further loss until the early appearance of senescence. However, the use of several methods of assessment at different ages makes it difficult to be certain that there is real progressive deterioration of intellect after birth.

Stern: Some of the progressive fall in scoring on intelligence tests may be artificial. With increasing age a higher standard of performance is required in order to attain the same level of I.Q score. If performance in the Down's syndrome infants remains the same, a progressively lower I.Q. score will be allocated.

Benson: I should like to comment on the need to distinguish between 5-hydroxytryptamine (5HT) and 5-hydroxyindole concentrations. For example the statement was made earlier that 5-hydroxytryptophan (5HTP) can produce a similar rise in the blood 5HT concentration in children with Down's syndrome as in controls. However, this does not appear to be the case. Bazelon *et al.* (1967 *Lancet* **i**, 1130) reported that treatment of subjects with Down's syndrome by 5HTP produced about a two-fold increase of the total blood of 5HI and that only a small proportion of this was due to a rise in 5HT. The final level of 5HT still being only about two-thirds of normal levels. Another example is the report by Dubowitz and Rogers, 1969, *Develop. Med. Child Neurol.* **11**, 730, who measured 5-OH indoles in the C.S.F. and found normal levels in Down's syndrome. However, although one can speculate, it is not possible from this study to say whether 5HT concentrations in the C.S.F. were normal or not.

RNA Synthesis by Down's Syndrome Leucocytes

P. F. BENSON

Paediatric Research Unit
Guy's Hospital Medical School, London SE1

It may be predicted that the expression of numerous genes carried on the supernumerary trisomic chromosome in cells from subjects with Down's syndrome is likely to be highly complex (Yielding, this symposium). Studies which have demonstrated biochemical differences in subjects with Down's syndrome and controls (Stern, this symposium) have not established any specific biochemical characteristic peculiar to Down's syndrome. It should be considered whether an additional G-group chromosome may be expressed phenotypically by causing changes in the rates of biological processes, for example protein and RNA biosynthesis (Benson, 1967). There is some evidence which suggests that nucleolar organisers are carried on G-group chromosomes (Polani, this symposium, p. 168). If redundancy of cistrons for ribosomal RNA were to produce an accelerated rate of ribosomal RNA synthesis, and if the number of ribosomes were rate-limiting, a net increase in the rate of protein synthesis might result. Previous studies have shown that leucocytes from subjects with Down's syndrome have higher rates of incorporation of ^{14}C-orotic acid and ^{3}H-uridine into RNA; and of ^{14}C-glycine into protein, than leucocytes from normal controls (Benson, 1967). Moreover, in individual subjects a positive relation was found between the activities of the leucocyte enzymes glucose 6-phosphate dehydrogenase and alkaline phosphatase and the rates of isotopic precursor incorporation into leucocyte RNA and protein (Benson and de Jong, 1967). It seemed of interest, therefore, to study the rates of synthesis of different RNA types in a more homogenous population of cells. The studies described below were carried out on cultured lymphocytes with or without phyto-haemagglutinin stimulation.

TABLE 13.1

Incorporation of ^{3}H-uridine into RNA by cultured lymphocytes derived from children with Down's syndrome or from matched controls

Duration of incubation	*PHA*	*Controls*			*Down's syndrome*			*t*	*p*
		No. of subjects	*counts/min/10⁶ lymphocytes*		*No. of subjects*	*counts/min/10⁶ lymphocytes*			
			Mean	*S.D.*		*Mean*	*S.D.*		
1 hour	None	15	84.9	48.4	15	144.7	100.4	2.08	0.02–0.05
	1*	15	308.4	248.1	15	449.2	369.5	1.23	0.2 –0.3
	1/10	15	180.4	144.5	15	227.7	275.8	0.827	0.3 –0.4
	1/100	15	92.7	75.4	15	196.3	116.7	2.89	0.001–0.01
24 hours	None	18	541.5	422.9	15	2155.0	1162.0	5.49	0.001–0.01
	1*	18	5410.0	7133.0	15	7771.0	5159.0	1.10	0.2 –0.3
	1/10	18	1490.0	2033.0	15	5096.0	6078.0	2.37	0.02–0.05
	1/100	18	635.4	551.5	15	3323.0	2303.0	4.80	0.001–0.01

Lymphocytes were isolated and cultured (1.5×10^{6} lymphocytes/ml) as described in the text. They were incubated with or without different concentrations of PHA as indicated. After 18 hours ^{3}H-uridine (3μcu/ml) was added. One or 24 hours later RNA was extracted and assayed for radioactivity.

* Full strength PHA was 0.02 ml/ml of culture medium.

Incorporation of ^{3}H-uridine into RNA

Lymphocytes from 20 ml of heparinised blood from 18 children with primary trisomic Down's syndrome (confirmed by karyotype analysis) and 18 controls matched for sex and age were isolated and cultured in medium TC 199 containing 30% pooled human serum at a concentration of 1.5×10^6 lymphocytes per ml of medium (Benson, 1971). Some cultures contained phytohaemagglutinin (Wellcome Laboratories) (PHA) at concentrations shown in Table 13.1. After 18 hours ^{3}H-uridine (3μcu/ml) was added to all cultures. After either 1 or 24 hours, lymphocytes were counted and RNA was extracted and assayed for radioactivity (Benson, 1971).

The rates of incorporation of ^{3}H-uridine into RNA were significantly higher in lymphocytes derived from children with Down's syndrome than those from controls after both 1- and 24-hour labelling periods in cultures either without PHA or with PHA at 1/100 dilution (Table 13.1). The mean radioactivity of RNA (expressed as counts per minute/10^6 lymphocytes) was also higher in Down's syndrome lymphocytes in all other groups studied, but was significant at the 5% level only in cultures incubated with isotope for 24 hours and containing PHA at 1/10 dilution.

Sucrose Density Gradient Analysis of RNA Types

Lymphocytes were isolated, cultured and incubated with ^{3}H-uridine as described above. At the times after addition of isotope shown in Figs. 1 and 2 cells were counted and RNA extracted (cold phenol/SDS) and analysed by sucrose density centrifugation (5-20%) (Benson, 1971). In each case the position of 18s and 28s RNA was identified by continuous optical scanning of added rat liver ribosomal RNA marker. It will be seen in Figs. 1 and 2 that labelling of ribosomal RNA (45s precursor, 28s and 18s) was markedly more rapid in Down's syndrome than in control lymphocytes at all labelling periods, but especially after one hour. Radioactivity of RNA with polydisperse sedimentation characteristics was also higher in Down's syndrome than in controls.

Nucleotide Composition

Nucleotide compositions were determined (Katz and Comb, 1963) of radioactive lymphocyte RNA after incubation with $H_3{}^{32}PO_4$ (3 mc/ml) for 30 and 60 mins after preincubation with PHA (diluted 1 in 10) for 18 hours, in three children with Down's syndrome and in

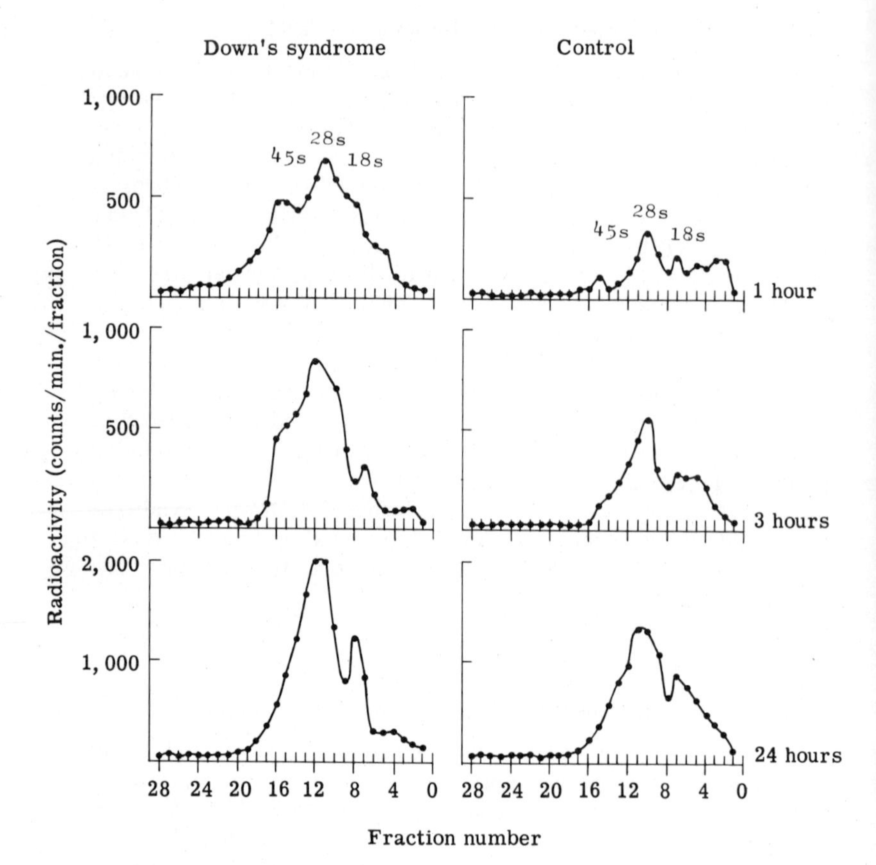

Fig. 1. Sedimentation patterns of radioactive RNA from lymphocytes cultured 18 hours with PHA (1/100); at progressively longer periods after the addition of ^{3}H-uridine (3μcu/ml). Centrifugation in 5-20% sucrose density gradient was for 4 hours at 105,000 *g* and 2°.

Fig. 2. Sedimentation patterns of radioactive RNA from lymphocytes cultured 18 hours with PHA (1/10); 1 hour after the addition of ^{3}H-uridine (3μcu/ml). Centrifugation in 5-20% sucrose density gradient was for 4 hours at 105,000 *g* and 2°.

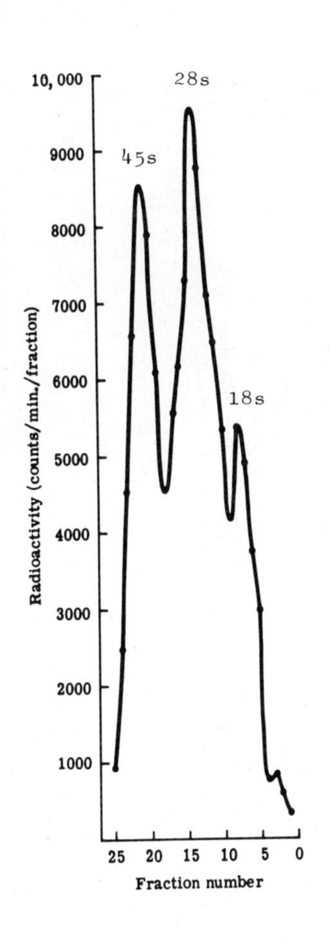

45s
28s
18s
Radioactivity (counts/min./fraction)
10,000
9000
8000
7000
6000
5000
4000
3000
2000
1000
25 20 15 10 5 0
Fraction number

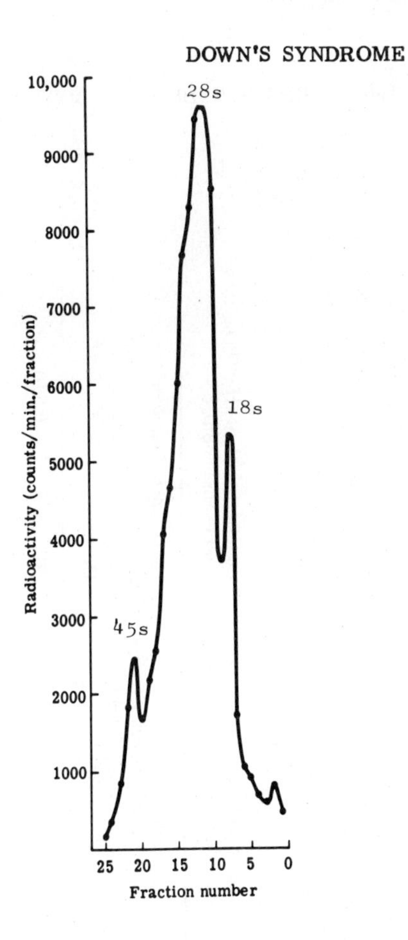

DOWN'S SYNDROME
28s
18s
45s
Radioactivity (counts/min./fraction)
10,000
9000
8000
7000
6000
5000
4000
3000
2000
1000
25 20 15 10 5 0
Fraction number

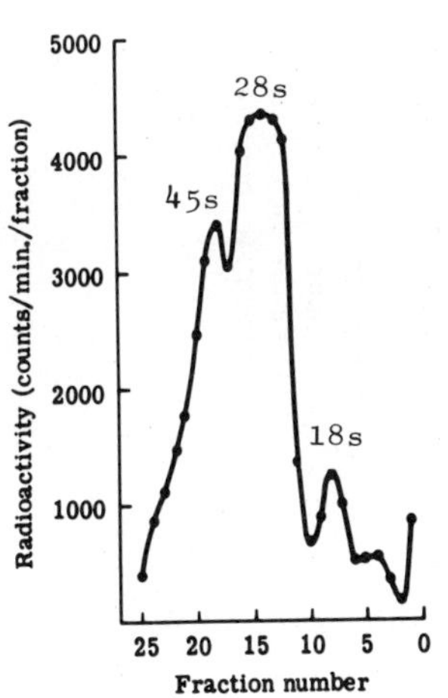

28s
45s
18s
Radioactivity (counts/min./fraction)
5000
4000
3000
2000
1000
25 20 15 10 5 0
Fraction number

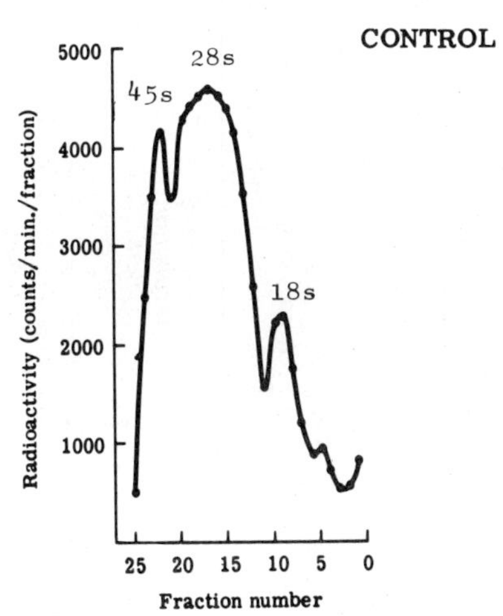

CONTROL
28s
45s
18s
Radioactivity (counts/min./fraction)
5000
4000
3000
2000
1000
25 20 15 10 5 0
Fraction number

TABLE 13.2

Mean nucleotide compositions and ratios determined by the incorporation of $H_3{}^{32}PO_4$ into lymphocyte RNA in three subjects with Down's syndrome and in three controls

Duration of incubation with $H_3{}^{32}PO_4$	*Source of lymphocyte RNA*	*Mean nucleotide composition (moles/100 moles)*				*Mean nucleotide ratios*
		UMP	*GMP*	*AMP*	*CMP*	$\frac{GMP + CMP}{AMP + UMP}$
30 min	Control	26.83	27.03	23.77	22.40	0.976
	Down's Syndrome	27.03	28.60	22.20	22.20	1.032
60 min	Control	26.10	30.03	19.93	23.93	1.174
	Down's Syndrome	21.33	34.20	19.63	24.93	1.445

Lymphocytes were cultured for 18 hours in the presence of PHA diluted 1 in 10 as described in the text before the addition of $H_3{}^{32}PO_4$ (3 mc/ml). After a further 30 or 60 min RNA was extracted and its nucleotide composition determined.

three controls. After 30 mins of incubation with isotope (Table 13.2), no convincing difference was observed between the results in subjects with Down's syndrome and those of controls. In each case the GMP + CMP/AMP + UMP molar ratios were close to unity and therefore unlike that of ribosomal RNA (Cooper, 1968). However, after 60 mins incubation with isotope the GMP + CMP/AMP + UMP ratios were higher in the three subjects with Down's syndrome than in the controls. This is in accord with the data from sucrose density gradient studies described above which indicate that the rate of incorporation of isotopic precursor into ribosomal RNA is greater in lymphocytes derived from children with Down's syndrome, than in controls.

The greater difference between incorporation rates of ^{3}H-uridine by Down's syndrome lymphocytes and control lymphocytes when stimulated by dilute rather than by full strength PHA is in accord with the observations by Hayakawa *et al.* (1968). These authors observed that a significantly higher proportion of lymphocytes from Down's syndrome than from controls were transformed and found to incorporate tritiated thymidine into DNA when 1/100 dilution of PHA was added to the culture medium, but not when full strength PHA was used.

In theory, the enhanced incorporation of ^{3}H-uridine by Down's syndrome lymphocytes could result either from an accelerated rate of transcription of the normal complement of cistrons or from average rates of transcription of an increased number of DNA copies such as might occur if the relevant cistrons were carried on the additional G-group chromosome and if these genes were not subject to tight regulatory restriction.

ACKNOWLEDGEMENT

I am grateful to the Spastics Society, The Wellcome Foundation and The Children's Research Fund for financial support.

REFERENCES

Benson, P. F. (1967). *Nature* **215,** 1290.

Benson, P. F. and de Jong, M. (1967). *Biochem. J.* **105,** 11P.

Benson, P. F. (1971). RNA and protein synthesis by Down's syndrome leucocytes. M.D. Thesis, London University.

Cooper, H. L. (1968). *J. biol. Chem.* **243,** 34.

Hayakawa, H., Matsui, I., Higurashi, M. and Kobayashi, N. (1968). *Lancet* **i,** 96.

Katz, S. and Comb, D. G. (1963). *J. biol. Chem.* **238,** 3065.

DISCUSSION

Allison: Have you considered that after PHA stimulation the normal cells may have got off to a flying start so that the G-trisomic cells may have appeared to synthesise RNA more rapidly than controls because the maximal rate of synthesis by the control cells was already over?

One should also look at the early events after PHA stimulation. These might be of interest because they would give information about rates of transcription. In your experiments results were related to the number of lymphocytes. One should be aware that the size of the lymphocytes in Down's syndrome may be abnormal. This would influence the size of soluble pools and a number of other factors.

Benson: Even with no PHA stimulation, during one hour of incubation, the G-trisomic lymphocytes incorporated ^{3}H-uridine into RNA more rapidly than control lymphocytes. Under these conditions (when almost all newly synthesised RNA is polydisperse,) the possibility of a greater initial incorporation into RNA by control cells than G-trisomic cells seems unlikely. Stimulation of RNA synthesis in normal cells by PHA can be detected within a few hours, but the peak mitotic activity and the maximal rates of RNA synthesis do not occur until 60-72 hours after incubation with PHA (Cooper, 1968, *J. biol. Chem.* **243**, 34). This is not compatible with the suggestion that maximal rates of synthesis by control cells was nearly over only 18 hours after incubation.

With regard to the expression of results, even if incorporation is expressed in units of specific activity, that is d.p.m/mg RNA, the results still indicate significantly more rapid rates of incorporation by Down's syndrome lymphocytes. This suggests that the differences in rates of ^{3}H-uridine incorporation are independent of cell size.

Allison: Another point is that taking the whole population of cells together, one cannot be certain that one is not missing changes produced by a few individual cells.

Yielding: There is some evidence that actinomycin can enhance the rate of RNA removal. It would be interesting to study RNA turnover in G-trisomics in the absence of actinomycin.

Polani: Dr. Benson quoted me as having suggested that the effect of trisomy '21' might be related to the presence of an additional chromosomal nucleolar organiser site, as defined by McClintock (McClintock, 1934, *Z.Zellforsch.* **21**, 294) and as taken to be the location for the reiterated ribosomal RNA genes (Wallace and Birnstiel, 1966, *Biochim. biophys. Acta* **114**, 296). Perhaps I could briefly discuss this. The effect of supernumerary chromosomes on genetic expression could be considered under three main headings (Polani, 1968, *Guy's Hosp. Rep.* **117**, 323; Polani, 1970, *Nature* **223**, 680). First, the effect could be specific, that is due to genes carried on the supernumerary chromosome. Secondly, the effect could be non-specific, i.e. due to the mere presence of an extra chromosome. This effect might be expressed possibly by the slowing down of the cell cycle. And thirdly, the effect may be semi-specific, i.e. an effect which might be produced by similar genes carried by several groups of chromosomes, for example those that carry a nucleolar organiser. This effect might be produced where there is a trisomy of either chromosome 21 or 13. The evidence for nucleolar organisers on these chromosomes is as follows. We were impressed right from the beginning of our studies on chromosome anomalies by the fact that so often the acrocentric chromosomes of man at metaphase seemed glued together (Polani *et al.*, 1960, *Lancet* **i**, 721). It is worthy that, when there is interchange of material from one chromosome to another, it is often these acrocentric chromosomes that are affected. Moreover, in man acrocentric chromosomes are often satellited and this possibly indicates that they bear nucleolar organisers. The data that have appeared subsequently by Harnden (Harnden, 1961, In: 'Recent Advances in Human Genetics' (ed. L. S. Penrose), p. 19. J. and A. Churchill, London) and the quantitative studies of Ferguson-Smith (Ferguson-Smith and Handmaker, 1961, *Lancet* **i**, 638; *Ibid* **ii**, 1362) and the semiquantitative work of Ohno and collaborators (Ohno *et al.*, 1961, *Lancet* **ii**, 123) all tend to confirm this.

Allison: You have shown that under the conditions of your experiments there is an enhanced rate of ribosomal RNA synthesis. What effect do you think this would have?

Benson: Control of protein synthesis can be exerted at a number of levels and involves many regulatory elements. Restriction on the rate of protein synthesis can be exerted when one or more of these is rate-limiting. There is evidence which suggests that under certain circumstances the amount of ribosomal RNA may be limiting in this way. For example, an enhanced rate of ribosomal RNA synthesis is a constant finding when the rate of protein synthesis is increased, in target organs after hormonal stimulation, in regenerating rat liver after partial hepatectomy, in rat liver during re-feeding after starvation and during the period of postnatal enzyme development.

Polani: Ribosomal DNA in humans comprises a relatively small proportion of the genome, that is about 0.008%, as compared with *Drosophila* and *E. coli* where it is about 0.3% and *Xenopus* with 0.10% (Brown, 1967, *Curr. Top. develop. Biol.* 2, 47). There are practically no data on the biochemical expression of supernumerary nucleolar organisers and only a few studies relating to deficiency nucleolar organisers.

Rees: In plants it is very easy to make trisomics of chromosomes with nucleolar organisers. Lin (1955, *Chromosome* 7, 340) some years ago working with maize found an almost linear relation between the quantity of RNA and the number of nucleolar organisers which varied from 1 to 4. The activity of the nucleolar organiser locus was quite independent of any other loci on the chromosomes.

Jain (1969, *Heredity* **24**, 59) working with the large chromosomes of hyacinths, also found that the amounts of cellular RNA were increased in plants trisomic for the nucleolar organiser chromosome.

I think there are other chromosomes which act in very different ways in relation to RNA content. We have observed that where the number of chromosomes is increased from one to two, there is an initial boosting of RNA synthesis; but when the number of these chromosomes is increased to more than three, there may be an inhibiting effect on RNA synthesis.

Polani: Yes, I think that these observations are very relevant to the human trisomic situation. One should not try to explain all the effects of trisomy on merely a supernumerary nucleolar organiser; other genes on the trisomic chromosome must also be considered.

Raine: Does mosaicism occur in metabolic diseases in humans?

Polani: It is easy to detect cellular mosaicism in chromosomal anomalies and it is also possible under other conditions, for example in RBC's; but it is not known whether this is developmental–that is whether it developed before embryogenesis was completed or whether it is proliferative–that is developed in proliferating tissue such as bone marrow, after embryogenesis was completed.

Raine: What percentage of individuals are mosaics?

Polani: Of Down's syndrome subjects, at least 1% are mosaics but this figure for development chromosome mosaicism may not apply to normal individuals. The proportion of sex-chromosome mosaics among newborn is also unknown but appears to be not negligible (on mosaicism in general see Stern, 1968, 'Genetic Mosaics and Other Essays'. Harvard University Press, Cambridge, Mass.). In ageing normal people, especially females, proliferative mosaics seem to occur by the appearance of presumptive XO (45,X) cells among the 46,XX lymphocytes (Court Brown *et al.*, 1966, Chromosome studies on adults, *Eugen. Lab. Memoirs* **42**, Cambridge University Press, London and New York; Hamerton *et al.*, 1965, *Nature* **206**, 1232).

Muir: I should like to ask Professor Yielding what effect would be anticipated in cells which reach the 'steady state' earlier than normal?

Yielding: This would depend on the kinetics of how the steady state is reached; if some cells reach steady state sooner than their neighbours, and assuming that the parameter we are considering (for example ribosomal RNA) is not able to function until the steady state is

reached, then these cells could get into trouble if they were out of step.
Polani: This may be very relevant to the human trisomies. If the steady state is reached earlier than it should be, so that, for example, some embryological stage is anticipated, the function of the relevant parameter may cut in earlier than it should do and upset the developmental balance.
Yielding: This effect can be visualised simply when considering the effects of stimulating precocious closure of osseous epiphyses by androgenic substances, where afer a growth spurt skeletal growth becomes arrested because of epiphyseal fusion and the final length of the bones is shorter than it would have been.
Polani: Another example which is relevant to RNA is the bobbed mutation in *Drosophila.* In the homozygous female this mutation can reduce body size if it is present during early development.
Allison: Are there any data on the amount of nucleolar RNA in cells of trisomies?
Polani: No there are not. Nor are there any good data on the number of nucleoli in these cells and this should be studied.
Rees: In plants one can do this using interference microscopy simply by weighing the nucleoli.
McIlwain: With respect to the point that has been made regarding the possible premature commitment of tissues in trisomies, is there evidence for subjects with Down's syndrome showing changes in adaptation either to drugs or to hypoglycaemia? These are two ways in which the adult brain can adapt.
Stern: We looked at this once. I found that mentally retarded children either with or without Down's syndrome had lower fasting blood sugars than normal children. Down's syndrome children also had flat glucose tolerance curves but they showed no clinical manifestations of hypoglycaemia. In this resistance to hypoglycaemia they resembled newborn infants more than controls of their own age groups. No change in their sensitivity to insulin has been observed. Two altered responses to drugs may be of interest. Following a suggestion by McKusick-Berg *et al.,* (1959, *Lancet* **2**, 441) tested children with Down's syndrome for the pupillary reaction to local atropine administration and found a much more sustained dilatation in Down's syndrome. Secondly, following oral aspirin administration children with Down's syndrome form less salicyluric acid than controls (Ebadi and Kuget, 1970, *Pediat. Res.* **4**, 187).
Polani: Norman (Norman, 1966, *Develop. Med. Child. Neurol.* **8**, 170) stressed that in brains of children with trisomy-D (Patau's syndrome) or trisomy-G (Down's syndrome) there was often cortical heterotopia; that is, the presence of masses of grey cells (usually poorly

differentiated) in ectopic sites, often in the region of the fourth ventricle. He suggested that these could be due to abnormalities of cell migration, i.e. migration to the wrong place, or to delayed migration. In relation to premature commitment possibly heterotopia might result from premature migration.

Cavanagh: In heterotopias if there is early migration of cells and if this results in the cells settling down earlier than normal, then cells might be stimulated to differentiate at an earlier stage than normal.

Muir: Are there any abnormalities in the E.E.G. pattern in Down's syndrome?

Polani: Yes. (Gibbs *et al.*, 1965, *Elektromedizin* **10**, 73; Martinelli *et al.*, 1964, *Riv. Neurobiol.* **10**, 539; Menolascino, 1965, *Amer. J. ment. Defic.* **69**, 653). This can be best explained on the basis of delayed maturation (Pampiglione, 1970, personal communication) and they are probably not related to the occurrence of heterotopias.

Yielding: The prolonged persistence of fetal haemoglobin in certain cases of Down's syndrome appears to support the concept of early commitment in these children.

Polani: There is also the persistence of the embryonic haemoglobins in trisomy 13.

Allison: I have been wondering to what extent abnormalities in Down's syndrome could be explained by prolonged mitotic interval; e.g. if in the development of the heart, rapid cell division is required at a critical stage, and if G-trisomic cells are unable to respond in this way they may develop morphological abnormalities. Also in the brain, delayed mitotic intervals might retard cellular migration and produce heterotopia. Is there any experimental evidence as to whether heterotopia could be produced if mitosis was delayed?

Cavanagh: I don't know of any but I think this could be the case.

The Effect of Chromosome Redundancy on Gene Expression

K. LEMONE YIELDING*

Professor of Biochemistry and Assoc. Professor of Medicine, Medical Center, University of Alabama, Birmingham, Alabama, 35233

The ultimate control of biological processes depends on the function of the genetic apparatus which serves to store information as well as modulate its expression into biological events. Thus, the system must be internally balanced and must be integrated appropriately with the cellular and external environments. It appears that there are also quantitative restrictions imposed on the genetic apparatus since the maintenance of the appropriate genetic load is critical to the survival of certain complex organisms. This has been emphasised in recent years by recognition of the devastating effects in humans of redundancy of specific chromosomes. A variety of differentiation and functional defects have been described, but the most common appears to be that of mental retardation. This is not surprising in view of the complexity of the differentiation and developmental processes in the central nervous system, but indeed may provide one of the few precise correlates between genetics and the development of the central nervous system.

At our present state of knowledge several important questions may be asked: First, what is the mechanism by which genetic redundancy leads to subnormal function? Secondly, what can be learned about normal genetic control and differentiation, particularly of the central nervous system, through study of trisomy states? Finally, what can be learned about such genetic abnormalities through study of lower organisms? The logical sequel to these questions are: What, if anything, can be done to correct or prevent trisomy-related defects; and what control or safeguards may we exert over normal differentiation?

In the present discussion I shall present several models involving

*Recipient of RCDA grant number GM 22698 from the National Institute of Health, Bethesda, Maryland.

regulation of genetic function which may serve to suggest the variety of effects which might be expected from genetic redundancy. These models are derived editorially from an interest in biological regulation based on the progress that has occurred over the past ten years or so in the understanding of genetic regulation in lower organisms. What is most apparent from such considerations is that redundancy may lead to a variety of effects rather than a simple increase in gene products from all the involved genes (Yielding, 1967). Those who deal with models are often accused of oversimplifying the problems of the more complex organisms. The extensive application of model systems in fact provides rather complex predictions about the possible behaviour of higher systems, although model testing is difficult. (It is interesting that the qualitative concept of '1 gene–1 enzyme' has so often been taken to be a quantitative concept, i.e. 1 n gene dose = 1 n level of enzyme . . . 2 n genes = 2 n level of enzyme.) The discussions to follow apply equally well toward understanding changes in gene ploidy and to predictions for the heterozygotic state. It must be remembered that normal genetic control and differentiation involves two important parameters: *Levels* of expression; and the *scheduling* or *staging* of gene function (time course) and its response to environmental signals. In the complexity of control and differentiation the latter may be quite difficult to describe or observe, but is, nonetheless, the key to differentiation and survival of the organism. Thus, disturbances may occur involving a combination of premature or delayed expression either at a subnormal or excessive level.

It is also important to emphasise that expression results from the complex inter-relations between the primary gene product and the environment, both internal and external. The importance of gene dosage is also determined by whether or not the gene in question controls a critical and rate limiting process.

EXPRESSION OF SUPERNUMERARY GENES IN MICROORGANISMS

Regulation of gene products in lower life forms as an aid to our understanding of the effect of ploidy on gene expression

The Repressor-operator Model and Transcriptional Control

It is clear that genes may be subject to negative control by *specific protein repressors* dictated by *specific regulatory genes* through binding to specific gene loci designated as *operators* (Jacob and Monod, 1961).

A number of repressor systems have been studied thoroughly in bacteria and evidence is accumulating that control depends on multisubunit proteins with specific binding sites for small *inducer* molecules which prevent repressor binding to the operator, probably through changes in repressor conformation. Thus, 'signal' evoked positive control occurs through release by inducer of the negative control by repressor. Small molecule evoked repression is explained by the same scheme if binding of the small molecule results in a conformation of the repressor which does bind to the operator. Studies on the repressor of the galactosidase operon in *E. coli* reveal it to be a multisubunit protein with a binding constant for inducer of the order of 10^{-13} M (Lieb, 1968; Riggs *et al.*, 1970). In this model, at least four parameters are under rather precise genetic dictation: The repressor (or regulator gene), operator structure, the relationship between operator and structure gene(s), and the levels of specific inducer. (The latter through control of its absorption, synthesis, and degradation.) Transcriptional control may be exerted by other specific mechanisms as well, particularly in higher organisms, but the usefulness of this well established model of transcriptional control for rationalising phenomena is well accepted.

Control of Translation

Considerable evidence is accumulating indicating complex mechanisms for translational control. First of all, translation depends on availability and efficiency of the different translational elements, i.e. tRNA, ribosomes, amino acids, activating enzymes, energy source, etc. Considerable opportunity also exists here for control of the synthesis of specific proteins, depending on what is the rate limiting parameter. For example, degeneracy of the coding mechanism allows for control of the ratios of various protein products through levels of specific tRNA's or their binding to the ribosome. This is illustrated beautifully by the various suppressor systems described in bacteria, involving either the synthesis of new tRNA species or distortion of the interaction between ribosomes and existing tRNA's. There is much current interest in the study of translational control, and this will, no doubt, be found an important aspect of genetic expression.

Control of Enzyme Activity Through Structural Control

The modification of protein function through changes in tertiary structure evoked by binding of specific small molecules to non-substrate sites is a mechanism of great importance in the regulation of biological activity (Tomkins and Yielding, 1961; Monod *et al.*, 1963; Yielding, 1970). This mechanism provides the basis for such phenomena as feedback control (Yielding and Tomkins, 1962) and hormonal and drug effects (Gerhart and Pardee, 1962) and illustrates that the *environment* of a protein molecule exerts discriminating control over its function. Since such effects are mediated through non-substrate sites through non-covalent binding, the effects may be positive or negative and freely reversible.

In addition to non-covalent ('allosteric') modification of protein activity, there are a variety of covalent changes in proteins such as adenylylation (Shapiro *et al.*, 1967) and phosphorylation (Krebs and Fischer, 1956) which produce changes both in function and in response to control. This is another complication of gene expression in which a particular enzyme (protein) function is under control of another gene site through the production of a specific enzyme. Clearly, the modifying enzyme may itself also be subject to control.

Control of Protein Stability or Turnover

In a complex organism it is rather naive to assume that the level of an enzyme is controlled only by its structural gene and by its repressor gene. In fact there is much evidence that enzyme turnover is an important parameter which must be considered, although relatively little is known about the precise mechanisms of control.

STRUCTURAL CONTROL OF STABILITY

In the absence of a specific protease for degradation of each enzyme, it may be assumed that the control of the stable configuration of a protein is of critical importance in determining turnover. It is well established that 'allosteric' effectors influence stability both negatively and positively, and thus may provide convenient means for regulating turnover (Grisolia, 1964; Bitensky *et al.*, 1965). Similarly, covalent modification of structure will influence stability. In this way, it is quite possible to modify levels of an enzyme in the absence of any change in its rate of synthesis. In addition, modification in the function of one pathway may change the levels of enzymes in another system which are under coordinating allosteric control. As an example of regulating protein stability through small molecule interactions the case of inducer stabilisation of the repressor of β-galactosidase may also be considered (Sadler and Novick, 1965).

OCCURRENCE OF SPECIFIC DEGRADATIVE ENZYME

The resulting complexity rules out that every enzyme should require a specific protease for turnover. Rechcigl has reported, however, a hereditable decrease in liver catalase in mice which is accounted for by a change in enzyme turnover, presumably from alteration of such a specific degradative mechanism (Rechcigl and Heston, 1967).

THE OCCURRENCE OF SPECIFIC 'BINDING' SUBUNITS FOR REGULATING THE STABILITY OF PROTEINS

Many proteins in their stable configurations exist as aggregates of several chains which may or may not be identical. In the case of non-identical chains, stability of one chain may be dictated by interchain interactions with partner subunits as has been proposed for the dissimilar chains of LDH isozymes by Fritz *et al.* (1970). In haemoglobin, the stability of the alpha chains appears to depend on the availability of beta chains for aggregation, as was pointed out by Dr. Allison in this conference. It may be that such a mechanism is of general importance.

What are the expectations for changes in gene functions resulting from gene redundancy based on the above control mechanism?

To answer this question, the role of the gene product must be known, whether this role is rate limiting in the cell economy, whether the cell is at steady state with respect to expression, and what controls are operating. For convenience of discussion, the models will be divided into several arbitrary categories.

Redundancy of 'Structure' Gene

The designation of only some genes as 'structure' genes is a matter of convenience. Any functioning gene of course yields a gene product which could be termed a 'structure'. However, it is convenient to classify those genes as structural which code for a specific measurable element of expression such as specific protein. There are obvious differences in predictions for changing the levels of degradative and synthetic enzymes, structural proteins, or protein subunits controlling the stability of other proteins.

UNCONTROLLED STRUCTURE GENES

In this simplest situation, in which protein synthesis is limited by messenger concentration and messenger synthesis is uncontrolled, a doubling of the gene dose should result in a doubling of the gene product ... assuming there is no concomitant change in product turnover. However, even in the absence of regulation at the transcriptional level it must be remembered that translational and post-translational control may be operating. In view of the complexity of higher organisms it even seems unlikely that in the simple case of doubling of uncontrolled genes a simple doubling of product would result. In fact, even in the *absence* of any control mechanism the question must be raised whether the cell is at steady state with respect to the product. Thus in the scheme: DNA → RNA → product → turnover, the rate at which a new steady state is achieved depends on the relative rates of these three processes in relation to the length of the cell cycle. In some situations steady state may be achieved only after many cell generations. (It is interesting to question whether part of the ageing process results from gene products which approach a steady state very slowly over a period of many cell generations).

CONTROLLED STRUCTURE GENES

Under this heading I should like to discuss several specific models in which the redundant structure gene is under varying degrees of transcriptional control by one of more regulatory genes.

Redundant structure gene plus redundant regulator (repressor) gene. The simplest model to consider is one in which the redundant chromosome contains both the structure gene and its relevant regulatory gene(s). Fortunately, this is a model for which we can examine data from bacteria. This model yields some rather surprising results. The question is whether increased, unchanged, or decreased enzyme levels should be expected. The answer, simply, is that any one of these results may be expected depending on the tightness of control.

In the case of tight control, one expects the structure gene product to vary inversely with repressor *concentration* (Gallant and Stapleton, 1963). A discussion of the β-galactosidase system in *E. coli* serves as a good example of tight control since recent studies by Riggs *et al.* indicate that a single molecule of repressor binds per structure gene sequence with an equilibrium constant of the order of 10^{-13} M (Riggs *et al.*, 1970). It is also clear from their studies that this repressor is a multisubunit protein. In this particular instance repressor synthesis is not regulated (Novick *et al.*, 1965) so that a double dose of the regulator gene should result in a doubling of the level of repressor at steady state. The final level of enzyme when *both* genes are duplicated depends first of all on whether and to what extent the structure gene is induced. The enzyme may be doubled, if fully induced; the same if a simple tight equilibrium between free and bound repressor is operative; or even decreased if the action of repressor is cooperative. Examination of data from the literature involving several variants of *E. coli* has shown that in fact there may be a substantial decrease in uninduced enzyme in some diploid cells while others are essentially unchanged from the haploid state (Yielding, 1967; Novick *et al.*, 1965). This principle was also illustrated in experiments by Anderson and Ogg (1966) who found that diploid *E. coli* cells produced in response to camphor had decreased levels of inducible enzymes in the absence of induction while constitutive enzymes were increased. Inducible enzymes could also be *induced* to a level of 2n expression in these cells, however. The finding that diploid cells may be more repressed than haploid cells may be explained either on cooperative binding of multiple repressor molecules (the occurrence of repressor in an equilibrium between 'active' repressor and its subunits), or on a decrease in the rate of repressor turnover. For example, if the level of active repressor is governed by an equilibrium between an aggregate and four inactive subunits, a doubling in subunit concentration could be expected to result in as much as a 16 fold increase in active repressor!

Redundant structure gene with normal repressor gene complement. It is not known at this stage of our knowledge whether or not gene loci concerned with regulation accompany the relevant structure gene on the same chromosome. If not, it would be possible to double the dose of a structure gene with the normal level of control. The extent to which gene expression would differ from the normal state would depend on the tightness of control and whether the system is in a state of repression or induction. Whether the kinetics of derepression would be altered would be determined by the relative affinities and number of interaction sites between repressor and structure gene, and between repressor and inducer.

Redundancy of Regulatory Genes

The question of gene redundancy can also be examined from the specific standpoint of the product of a gene which is primarily regulatory. In this discussion, attention must be directed at both negative and positive elements of control.

REDUNDANCY OF A NEGATIVE CONTROL ELEMENT (REPRESSOR)

As in the case of redundancy of 'structure' genes, this could exist both in controlled and uncontrolled systems. In bacteria, examples of repressors have been studied which appear to be inducible (Garen and Otsuji, 1964) or non-inducible (Gallant and Stapleton, 1963).

Uncontrolled repressor gene duplication would be expected to decrease expression of relevant structure genes. In the event these are on separate, unduplicated chromosomes, a reduction in function would be expected unless fully induced, and the extent of reduction would depend on the tightness of control. The case of redundancy of both repressor and structure genes has been discussed above.

Controlled repressor genes, when duplicated, could lead to some rather paradoxical results. The simplest case is that with both regulatory elements and the structural unit under their ultimate control placed on the same redundant chromosome. Thus, an excess of repressor #1 could lead to a 'deficiency' of repressor #2 with resulting increase in expression of the structure gene. In cases of such complex control with placement of the structural gene on a different chromosome the increase in structure gene function could lead to erroneous conclusions for gene mapping.

Redundancy of translational units. This could lead to increased expression of gene loci which might be located on the same or different chromosomes, the expression for which is limited by the translational process. In this regard Dr. Benson has suggested in this conference that cells with trisomy 21 may show over production of ribosomal RNA. It may be that many chromosomes carry genes for translational elements so that such over production could be a general feature of chromosomal redundancy. In addition to the quantitative aspects of translation there may also be a great deal of qualitative specificity, especially with respect to tRNA, leading to specific changes in gene expression. As an example of how a specific tRNA can lead to increased expression of a specific gene, one has only to consider the case of suppressor genes in bacteria. It has been shown clearly, for example with a specific amber mutant of *E. coli* K_{12} for the alkaline phosphatase gene, that functional enzyme will be produced only when the correct suppressor gene is also present to contribute the required 'abnormal' tRNA. Furthermore, cells diploid for the suppressor gene can produce twice as much enzyme. If, in fact, in the process of differentiation a whole family of events is switched on by the appropriate level of translational components, redundant function could provide a premature and excessive 'switching on'.

The additional question must be raised of whether specific regulatory units also exist for control of translation as has been proposed by Tomkins *et al.* (1969). In such an event, the effects of redundancy are even more complex.

In this discussion it has become apparent that the time course of gene expression is critical with any type of gene. Although it is most simple to consider steady state conditions, the changes of greatest interest may be those in early pre-steady state which for the differentiation process is long before birth. It will be of great interest to extend our knowledge more extensively into the pre-natal period.

There are also a number of models which could be studied in higher organisms. These include drug-induced changes in ploidy, cell fusion, cloning of cell lines from heteroploid organs, and the extended study of various developmental abnormalities. The selection of enzyme or other function for study is critical, of course, because of the various factors discussed. It may be of particular interest to study processes known to be under rather tight control. One such example may be seen in the work described by Schneyer (1962) and Schneyer *et al.* (1967) involving isoproterenol induced hyperploidy in the salivary glands. They observed a considerable decrease in level of amylase which returned to normal as the ploidy reverted to normal. The secretory enzyme of a gland might be expected to be under rather tight control.

Summary

1. Prediction of the results of chromosomal redundancy requires extensive knowledge of the nature of involved gene products, to what extent and how expression is controlled, whether rate limiting processes are concerned in the overall cell economy, and whether the system in question functions under steady state or pre-steady state conditions.

2. Based on consideration of model systems and regulatory mechanisms which are known, it may be predicted that *increased, unchanged,* or *depressed* expression may result at steady state from chromosomal redundancy. Even in the absence of changes achieved at steady state, the time course of the approach to steady state may be modified in a critical manner.

3. Although predictions are difficult at the present time, the continued study of gene expression as a function of gene dose may lead to elucidation of the developmental abnormalities as well as provide insight into the factors which regulate gene expression in higher life forms.

REFERENCES

Anderson, D. E. and Ogg, J. E. (1966). *Amer. Soc. Microbial. Bact. Proc.* **66,** 82 (abstract).

Bitensky, M. W., Yielding, K. L. and Tomkins, G. M. (1965). *J. Biol. Chem.* **240,** 1077.

Fritz, P. J., Pruitt, K. M., White, E. L. and Vesell, E. S. (1970). *Nature* in press.

Gallant, J. and Stapleton, R. (1963). *Proc. Nat. Acad. Sci.* **50,** 348.
Garen, A. and Otsuji, N. (1964). *J. Mol. Biol.* **8,** 841.
Gerhart, J. C. and Pardee, A. B. (1962). *J. Biol. Chem.* **237.** 891.
Grisolia, S. (1964). *Physiol. Rev.* **44,** 657.
Jacob, F. and Monod, F. (1961). *J. Mol. Biol.* **3,** 318.
Krebs, E. G. and Fischer, E. H. (1956). *Biochim. Biophys. Acta* **20,** 150.
Lieb, M. (1969). *J. Mol. Biol.* **39,** 379.
Monod, J., Changeux, J. P. and Jacob, F. (1963). *J. Mol. Biol.* **6,** 306.
Novick, A., McCoy, J. M. and Sadler, J. R. (1965). *J. Mol. Biol.* **12,** 328.
Rechcigl, M., Jr. and Heston, W. E. (1967). *Biochem. Biophys. Res. Commun.* **27,** 119.
Riggs, A. D., Suzuki, H., and Bourgeois, S. (1970). *J. Mol. Biol.* **48,** 67.
Sadler, J. R. and Novick, A. (1965) *J. Mol. Biol.* **12,** 305.
Schneyer, C. A. (1962). *Am. J. Physiol.* **203,** 232-236.
Schneyer, C. A., Finley, W. H. and Finley, S. C. (1967). *Proc. Soc. Exp. Biol. Med.* **125,** 722-728.
Shapiro, B. M., Kingdon, H. S. and Stadtman, E. R. (1967). *Proc. Nat. Acad. Sci.* **58,** 647.
Tomkins, G. M., Gelehrter, T. D., Granner, D., Martin, D. M., Jr., Samuels, H. H. and Thompson, E. B. (1969). *Science* **166,** 1474.
Tomkins, G. M. and Yielding, K. L. (1961). *Cd. Spr. Hb. Symp. Quant. Biol.* **26,** 331.
Yielding, K. L. (1967). *Nature* **214,** 613.
Yielding, K. L. (1970). In 'Enzyme Synthesis and Degradation'. Karger Co., 1970).
Yielding, K. L. and Tomkins, G. M. (1962). *Recent Progr. Hormone Res.* **18,** 467.

DISCUSSION

Allison: One thing I find confusing is that according to this analysis there is a large number of possibilities which allows almost any type of change in enzyme activity to follow chromosomal trisomy and it is difficult to make specific predictions which could be tested experimentally.

Yielding: Yes, I think this is true. Two practical points emerge, however. First, that one should be aware that trisomy could lead to a reduction in the synthesis of gene products rather than an increase; and second, that we have to know the nature of the gene product before we can study it or make any predictions.

Rees: There are certain situations where the effect of increasing gene dosage can be studied. For example the effect of duplication of the Bar gene in *Drosphila* upon the number of eye facets.

Yielding: It can also be tested in bacteria. Anderson and Ogg (1966, *Amer. Soc. Microbiol. Bact. Proc.* **66**, 82) grew *E. coli* in the presence of camphor and found that although DNA is synthesised, the organisms fail to divide. Under these conditions there was an increase in uncontrolled enzymes but a decrease in inducible enzymes.

Pollitt: If the expression of structural genes for enzymes in mammalian systems is subject to the degree of control which Professor Yielding has described for microorganisms, it is surprising that in a large number of inborn errors of metabolism, enzyme activity in heterozygotes is half that found in normal homozygotes and this does suggest that in these cases the enzyme activity is controlled directly by dosage of structural gene and there is very little regulation by substrate levels or the like.

Yielding: This is not incompatible with regulated systems. I described mainly control at the transcriptional level but enzyme activity may also be controlled by cytoplasmic factors; that is by factors other than the number of enzyme molecules; for example allosteric and other types of control.

Scriver: In terms of metabolic efficiency rather than enzyme activity it is surprising how often heterozygote carriers of abnormal genes are completely normal, so much so that one might have predicted that the presence of a so-called extra allele, in chromosomal trisomies, might also have been without visible effect.

Raine: A genetic consideration is that there are practically no inborn errors inherited in a dominant fashion presumably because, being enzyme deficiencies, the normal complement of enzyme activity is great enough for this to be halved (as in the heterozygote), without the amount of enzyme exerting a limiting effect on the reaction it determines.

Polani: The porphyrias are interesting exceptions to this rule, being inherited in a 'dominant' manner and persisting in the population; yet it is clear that these conditions are 'inborn errors', albeit metabolically obscure (Harris, 1970, 'The Principles of Human Biochemical Genetics'. North Holland Publishing Co., Amsterdam and London, p. 185).

Benson: Both types of control exist even in bacteria. Activity of constitutive enzymes is related to gene dosage. Reduction of enzyme activity due to increased dosage of regulatory genes is seen in inducible systems where the rate of transcription of structural genes is controlled by repressors.

Yielding: Yes, that is correct.

Polani: Another factor to be considered is the presence of a nuclear membrane in eukaryotic cells which may influence post-transcriptional control of protein synthesis.

Raine: Could one investigate the effect of gene dosage on enzyme activity by studying the effect of inactivation of the X-chromosome in females? One might expect to have halving of activity of *X*-linked enzymes.

Polani: I don't know of any direct evidence for this but Epstein, who worked with early mouse embryos, showed a reduced activity of *X*-linked G6PD in *XO* as compared with *XX* germ cells.
Milne: Certain similarities come to mind between the trisomies and certain morphological abnormalities with dominant inheritance in man. These include polyposis coli and polycystic kidney. In both types of disorder there are very obvious structural changes but no known specific biochemical abnormalites. To my mind establishing the link between genetic expression and morphological changes is just as important as searching for biochemical abnormalities.
Yielding: The fundamental question is to establish the nature and function of the gene products whether they are expressed morphologically or biochemically.
Scriver: I think this lack of tying up between biochemical and physical phenomena is universal, e.g. a single amino acid replacement responsible for the formation of Hb-S rather than Hb-A has been clearly identified. However, the relation between the primary amino acid structure and the precise cause of sickling still remains to be determined. Does Dr. Allison agree?
Allison: Well, we are getting fairly close to it. We know that replacement of a charged amino acid by an uncharged produces a hydrophobic site at which haemoglobin aggregation takes place when oxygen saturation is low.

Chromosome Gain in Higher Plants

H. REES AND R. N. JONES

Department of Agricultural Botany, University College of Wales, Aberystwyth

The first and perhaps the most important point to make about plants, especially the flowering plants or Angiosperms, is that they display a remarkable capacity or tolerance for quantitative change in chromosome material. This capacity is made abundantly clear when we compare the range of nuclear or chromosomal *DNA* variation found between Angiosperms with that found between species of other groups of living organisms. Figure 1 and Table 15.1 show nuclear *DNA* differences of the order of 100 fold between species of flowering plants. In contrast, we see that the range of nuclear *DNA* values between mammalian species is relatively trivial. There is, in respect of *DNA* quantity, if not quality, precious little difference between men and monkeys. Amongst the animals generally, the only group exhibiting anything like comparable tolerance to the flowering plants in nuclear *DNA* change is that of *Amphibia.*

There is more than one cause and, of course, more than one consequence of the chromosome variation. The purpose of this article is to survey the causes and, so far as possible, their physiological and genetical consequences. We shall attempt also, to ascertain as far as possible, why the higher plants display such large and widespread changes and why the higher animals do not.

CAUSES OF CHROMOSOME GAIN

The chromosome changes fall into four convenient categories,

(a) Changes involving whole sets of chromosomes, *viz.*, polyploidy.
(b) The addition of whole chromosomes but not of complete sets, *viz.*, aneuploidy.

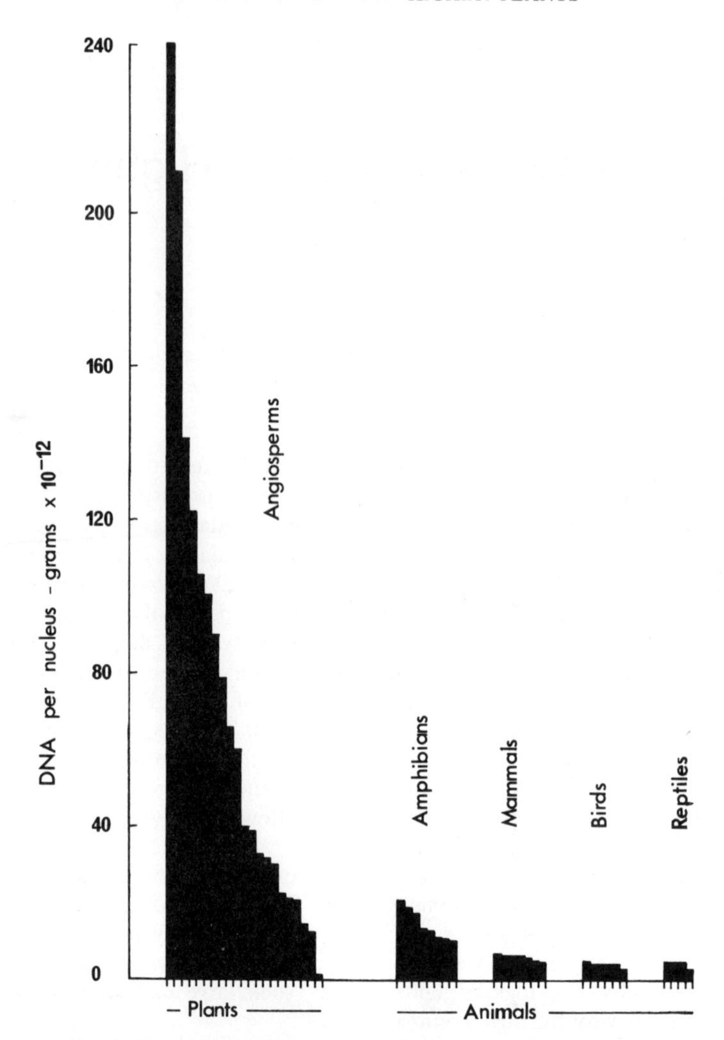

Fig. 1. The range of *DNA* variation in a variety of plant and animal species. Species within each group are arranged in descending order of *DNA* amount along the horizontal axis.

(c) Changes involving the generation of *new*, supernumerary or *B*-chromosomes.

(d) Change involving parts of chromosomes, including the amplification of localised segments.

TABLE 15.1

DNA values per 2*C* nucleus (in grams x 10^{-12}) for a variety of plant and animal species.

Plants		*Animals*	
Angiosperms		*Amphibians*	
Allium babingtonii	244.2	Proteus anguineus	21.1
Tradescantia ohioensis	211.0	Salamandra maculosa	18.4
Lilium longiflorum	141.1	Triturus cristatus	17.9
Allium ampeloprasum	122.2	Bombinator pachypus	13.7
Tradescantia ohioensis	104.8	Rana catesbiana	12.6
Tulipa gesneriana	100.7	Bufo calamita	10.7
Scilla campanulata	89.9	Pelobates fuscus	10.5
Allium cepa	78.4	Rana temporaria	10.3
Narcissus pseudonarcissum	64.9	*Mammals*	
Vicia faba	60.0	Rattus norvegicus	6.8
Galtonia candicans	40.0	Homo sapiens	6.5
Hepatica acutiloba	39.0	Sus scrofa domesticus	6.5
Phalaris minor	33.4	Ovis aries	6.5
Phalaris hybrid	32.0	Mus musculus	6.0
Zea mays	30.2	Bos taurus	5.6
Pisum satirum	22.5	Canis familiaris	5.5
Phalaris caerulescens	21.4	*Reptiles*	
Anemone riparia	21.0	Natrix	5.0
Agave attenuata	14.4	Chelydra serpentina	5.0
Pulsatilla occidentalis	12.2	Alligator	5.0
Aquilegia caer.xchrys.	1.3	Pseudechis porphyriacus	3.0
		Birds	
		Anser anser	4.9
		Anas platrhyncha	4.2
		Gallus gallus	4.2
		Phasianus colchicus	4.2
		Numida meleagris	4.2
		Cairima moschata	3.2

The kinds of nuclear accident which give rise to *a* and *b* are familiar enough and so, too, are the main consequences of these numerical chromosome changes. For these reasons we shall concentrate mainly on matters which influence their survival in populations. We shall deal at some length with *c* and *d* because recent information about them throws particular light upon the effects of 'redundant' *DNA* upon the activities of the nucleus, upon growth and development and upon the genctic systems of plant populations.

Polyploidy

Physiological

First, it is worth pointing out that in plants, as in most animals, a widespread partial polyploidy is a natural consequence of differentiation within normal diploids. Lewis (1967) has pointed out that a substantial proportion of both shoots and roots in many plant species, is composed of polyploid cells. And on these grounds he argues that any physiological advantages conferred by polyploidy upon individuals would readily be achieved by the induction and restriction of polyploidy to differentiated, non-dividing cells. Above all, this would avoid the complications at meiosis which give rise to the infertility characteristic of most polyploid organisms. Lewis goes on to argue that from the standpoint of adaptation we should therefore, look at and evaluate polyploidy not so much from a physiological point of view but from a different, genetic point of view; in terms of the effects of polyploidy upon the control and regulation of genetic variability within populations. We shall return to this genetic issue later. In the meantime, while there is no denying Lewis' argument in general terms there is, at the same time, no doubt that certain physiological features are common and characteristic of polyploid plants. Compared with diploids they exhibit,

1. Slower initial growth and later maturing.
2. Greater tolerance to extremes of temperature and rainfall.
3. Thicker more leathery leaves.
4. Greater size.

These immediate consequences of polyploidy may partly account for their superiority as colonisers in comparison with their diploid ancestors (Darlington, 1956). A clear example is provided by the Rhododendrons. In southern Asia the higher polyploids, 12x and 8x, are found in regions above 8,000 feet. Tetraploids and hexaploids extend from about 1,000-8000 feet, while the diploids are mainly confined to the lower land (Janaki Ammal, 1950).

One must hasten to add that the physiological and morphological changes we have enumerated by no means apply to all polyploids in all species. In the *Tradescantias* for example, diploid and tetraploid forms are indistinguishable from each other externally (Swanson, 1960). Even within species the consequences of polyploidy vary between different genotypes. In this connection, tetraploids produced from different inbred homozygous lines in rye generally show an increase in plant weight over the diploids. The increase is, however, greater for some lines than others (Fig. 2). In brief, the physiological effects to which we have referred are common but by no means inevitable.

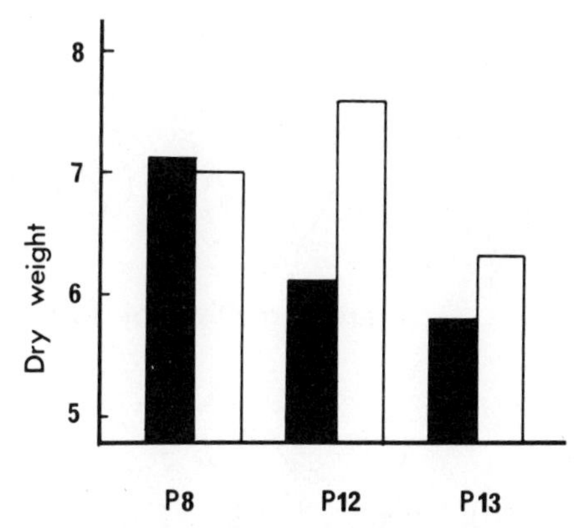

Fig. 2. Mean dry weights, in grams, of three diploid inbred lines of rye, *P8, P12, P13* and their induced autotetraploids.
■ = 2X □ = 4X. (Unpublished data kindly supplied by J. M. Hutto.)

This is not the place to attempt to explain how these effects are brought about. There are many contributory causes (Stebbins, 1966). Two obvious ones are, (a), the increase in cell size associated with increase in nuclear mass and, (b), the probable increase in enzyme production consequent upon increasing gene dosage.

Cytological

Polyploidy creates no problems to the efficiency of mitotic divisions. As for meiosis all odd numbered polyploids are characteristically incompetent. Segregation is irregular and, as a result, there is considerable, sometimes complete infertility. Turning to even numbered polyploids, we must of course, distinguish sharply between *allopolyploids* which are derived from hybrids between different species and *autopolyploids* which derive from diploids within the one species. In the former, association is confined to pairs of homologous chromosomes at pachytene of meiosis and to bivalents at first metaphase. Disjunction is regular and the fertility high (see Fig. 3). In autopolyploids multiple association of chromosomes at pachytene give rise to multivalents at first metaphase with consequent irregular disjunction and an inevitable, sometimes severe loss of fertility. It is worth pointing out that the degree of infertility in autopolyploids is to some extent under genetic or genotypic control (Hazarika and Rees, 1967) and a selection for improved fertility is effective in improving the fertility by achieving a more regular disjunction of chromosomes at meiosis (see Crowley and Rees, 1968).

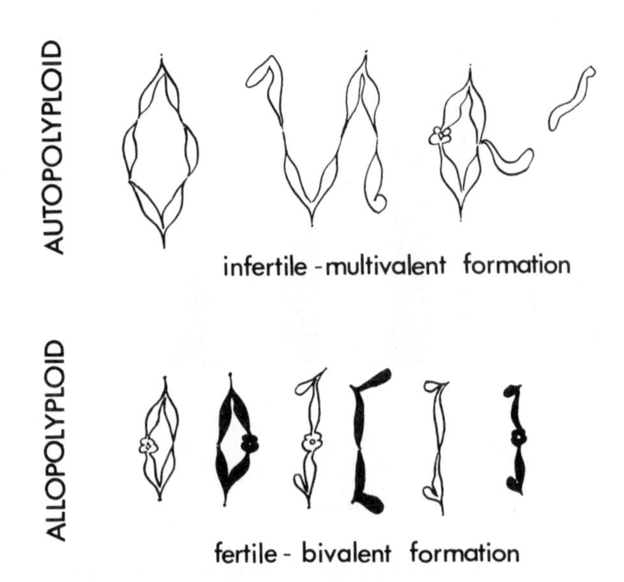

Fig. 3. Fertility and chromosome pairing at meiosis in auto- and allopolyploids.

Distribution and Variation

Despite the obvious hazards of polyploidy, particularly of autopolyploidy to fertility it is clear that this type of numerical change has played a notable part in the evolution of higher plant species. It has been estimated that more than 50% of Angiosperms are polyploid or of polyploid ancestry (Stebbins, 1963). It is widely distributed throughout this group, although it is of interest to note some sharp differences between genera of close affinity. Within the grasses for example polyploidy is common in *Avena, Triticum, Festuca,* and *Dactylis;* absent or rare in *Zea, Secale, Hordeum* and *Lolium,* even though polyploidy is readily induced in these last four.

Polyploidy in many cases clearly confers advantages which outweigh the, often, severe reduction in fertility. In this context it has been shown that polyploids, *under certain circumstances,* outyield diploids in mixed diploid/tetraploid populations. An illustration of this is given by Norrington-Davies and Crowley (1969). Induced tetraploids of *Lolium perenne* outyield the competing diploids under all conditions of spacing and fertility, except high fertility and low density. This *Lolium* work emphasises that in assessing by experiment the relative performance of polyploids and diploids it is certainly important to do so under competitive conditions over a wide range of cultural procedures.

To return to the theme of Lewis (loc. cit.), that the adaptive significance of polyploidy lies mainly in its effect upon the control of genetic variability, Lewis lists the following points,

1. Allopolyploidy serves to maintain the heterozygosity derived from the hybridisation of different species.
2. Even in autopolyploids the proportion of heterozygotes in the population is increased in comparison with the diploids (see Table 15.2).

TABLE 15.2

The relative frequencies of heterozygotes and homozygotes in populations of diploids, autotetraploids and allotetraploids: assuming a 1 : 1 distribution of alleles (*A* and *a*), random mating and no selection

		Heterozygotes
Diploid	AA 1 Aa 2 aa 1	50 per cent
Autotetraploids	AAAA 1 AAAa 4 AAaa 6 Aaaa 4 aaaa 1	87.5 per cent
Allotetraploid	AAaa 1	100 per cent

Heterozygous advantage or heterosis is, of course, well established in most outbreeding populations and it may well be as Lewis maintains that the generation and maintenance of heterozygosity is the chief asset resulting from polyploidy.

Aneuploidy

Because polyploidy is so common in plants it is not surprising to find that aneuploids, also, turn up far more frequently than in higher animals. The reason is, of course, that multivalent formation gives rise to irregular disjunction of chromosomes at meiosis in polyploids and, as a consequence, to aneuploid gametes and to aneuploids amongst the progenies. Amongst the progenies of autotetraploids we expect to find quite a few individuals with chromosomes in excess or in deficit of the 4x value. In *Lolium perenne* the proportion is from about 9 to 30% (Ahloowalia, 1970). Even under experimental conditions, however, only a limited fraction of aneuploids survive either as gametes or zygotes. Not surprisingly it is that group with chromosome numbers near to the euploid values. We note that 4x–1 and 4x+1 occur more frequently than aneuploids which result from the loss or gain of more than one chromosome. This is not only because loss and gain of one or few chromosomes at meiosis are more frequent but also because the survival of near euploid gametes and zygotes is superior.

In a comparable way the addition of a single chromosome to the diploid number to give a trisomic (2x+1) is less disastrous by far than the addition of two or more chromosomes. Put another way, the less the departure from the euploid value the less the degree of genetic imbalance (see Darlington and Mather, 1949). This is not to say, however, that trisomics are free of the symptoms of genetic imbalance. On the contrary they are almost invariably abnormal in morphology and frequently highly infertile.

Detailed work on the familiar tomato, (Rick and Barton, 1954) makes the point forcibly enough. The tomato work is instructive from another standpoint. In Table 15.3 are the frequencies with which

TABLE 15.3

The relative frequencies of the 12 *trisomics* in the progenies of triploid tomatoes

Chromosome number	*Frequency of* trisomics *as percentage of total progeny*	*Chromosome length in microns*
1	0.5	52.0
2	3.3	42.1
3	1.6	40.3
4	9.9	35.0
5	6.3	31.4
6	0.7	31.3
7	3.5	27.5
8	4.6	27.5
9	2.7	26.8
10	5.5	25.7
11	0.1	24.8
12	4.0	22.5

different kinds of trisomics turned up in the progenies of triploids. It will be observed that some trisomics are more frequent, more viable as gametes or zygotes than others. Trisomics for chromosomes 4 and 5 are relatively numerous in comparison with plants trisomic for 1 and 11. It will also be observed that the most common trisomics are not those involving the smallest chromosomes of the complement.

Evidently the size of the chromosome, the amount of genetic material it carries, is no certain guide to the extent of imbalance which it generates. This, of course, is just as true for humans where the most common autosomal trisomics are trisomic for 21 and although 21 is one of the smaller it is, after all, not the smallest chromosome.

There are species of higher plants which are exceptional in showing tolerance to aneuploidy. The best studied is the common Hyacinth, *Hyacinthus orientalis* (Darlington *et al.,* 1951) in which a whole range of aneuploids are effective and viable. The garden varieties of Hyacinths comprise not only diploids 2x = 16, triploids 3x = 24, and tetraploids 4x = 32 but, as well, aneuploids with 17, 19, 20, 21, 23, etc. Even here however, some aneuploid chromosome combinations are more common, more 'effective' than others.

We must emphasise that species like *Hyacinthus orientalis* are quite exceptional.

From the adaptive and evolutionary standpoint, aneuploids in higher plants are of little consequence. The next, less familiar and somewhat neglected category of numerical change is that involving supernumerary or *B*-chromosomes.

B-chromosomes

B-chromosomes, in contrast to aneuploids, are widespread in natural populations and the source of unusual, even surprising but clearly adaptive variation in the phenotype of individuals and in the genetic system of populations. The occurrence of *B*-chromosomes is recorded in some two hundered species of flowering plants (Battaglia, 1964), and they are especially common among members of the *Gramineae.* In general they are rare in animals but common in some groups, for example Grasshoppers (John and Hewitt, 1965).

The special feature of *B*-chromosomes which distinguishes and differentiates them from the normal *A*-chromosome complement, is the fact that their presence is not essential for the normal processes of growth and development. They are dispensable and apparently redundant. At the same time their widespread distribution throughout the plant kingdom, coupled with the fact that within a species their frequency varies between populations of different origin and habitat, suggests that they are nevertheless of some adaptive significance.

It is worth while, at this juncture, considering the distribution pattern of *B*-chromosomes in more detail, with reference to one particular species, namely rye (*Secale cereale*) in which they have been most intensively studied. In Table 15.4 are listed the frequency of occurrence of plants with various numbers of *B*'s, representing different

TABLE 15.4

B-Chromosome distribution in rye populations from various geographical locations

B-class	*0*	*1*	*2*	*3*	*4*	*5*	*6*
Iran*	1,812	14	137	1	4	–	–
Korea*	103	8	92	4	22	–	–
Japan†	6	1	21	4	26	–	5

* Muntzing, 1957
† Kishikawa, 1965

populations. As will be seen from the table, rye plants may carry a range of numbers (0-8) of *B*'s, and their distribution pattern varies for the different populations. In general they occur most frequently in the less well cultivated varieties. One thing however is common to all the populations, the preponderance of plants with even numbers of *B*'s as compared with odd numbers.

Inheritance

The excess of even numbers is explained by the mechanism of their inheritance which is non-Mendelian (see Hasegawa, 1934; Muntzing, 1946; Hakansson, 1948; Battaglia, 1964). At the first mitosis following meiosis there is a non-disjunction of *B*-chromosomes at anaphase such that both daughter chromatids of each *B* move to one pole. Functional sperm and egg nuclei tend to contain, in consequence, pairs of *B*'s rather than single *B*'s or other odd numbered combinations. This mechanism of inheritance tends also towards the numerical increase of *B*'s in the population, a tendency which is countered by the infertility of plants with high *B* frequencies.

The origin of *B*'s is obscure. In rye, as in most other species, they are smaller than the *A*-chromosomes, they are heterochromatic, replicate their *DNA* later in interphase than do the *A*'s. They show no homology with any members of the *A*-chromosome complement.

Effect upon Growth and Fertility

The presence of *B*-chromosomes represents, of course, a gain in the genetic material of the nucleus. One striking feature of this genetic material is that it carries no genes with major effects, in other words no major gene mutations are located in the *B*-chromosomes. This is not to say however that they are genetically inert. On the contrary they do cause variation, of a continuous nature much like that due to polygenes, in a wide range of characters affecting growth and development.

TABLE 15.5
Fertility, expressed as percentage seed set, in rye plants with 0-8 *B*-Chromosomes

B-class	*0*	*1*	*2*	*3*	*4*	*5*	*6*	*7*	*8*
% seed set	49.5	31.4	34.2	21.5	5.1	7.1	1.7	0.1	0

The most striking consequence of the genetic activity of *B*-chromosomes is upon fertility. An example is provided in Table 15.5, which gives the percentage seed set in rye plants with different numbers of *B*-chromosomes. It will be seen that with high numbers the plants are almost completely sterile, and even with a low *B* frequency their reproductive capacity is considerably reduced. This infertility is partly, but only partly, explained by the influence of *B*-chromosomes upon the behaviour of *A*-chromosomes at meiosis. The presence of many *B*'s is a cause of failure in chiasma formation in some *A*-chromosomes and, consequently, of univalent formation and chromosome loss (Jones and Rees, 1967).

The effect of *B*'s upon growth as reflected by plant height and weight is, in general, deleterious, particularly where the *B* frequency is high (see Fig. 4). This figure shows also a remarkable difference between the effects of *B*'s upon the phenotype in even as compared with odd numbered combinations. *B*'s in odd numbers, 1, 3 and 5, have a disproportionately deleterious effect upon the phenotype. We have argued that the non-Mendelian inheritance which assures a preponderance of even numbered *B*'s in the population is an adaptive device which serves to reduce the more deleterious consequences with populations attributable to *B*'s in odd numbered combinations (Jones and Rees, 1969). As Table 15.4 has shown the frequencies of plants with odd numbers of *B*'s are as expected disproportionately low relative to evens.

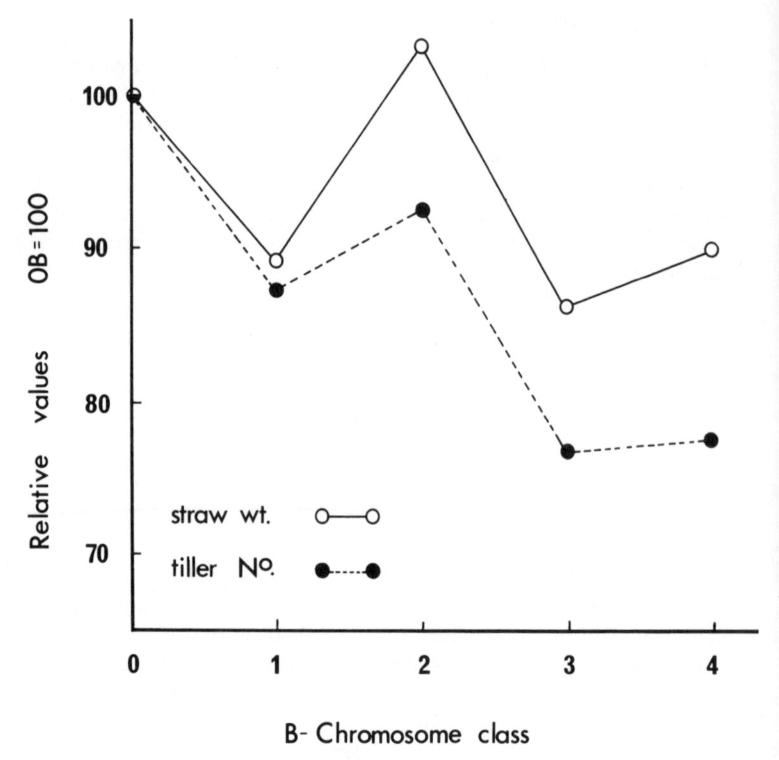

Fig. 4. Straw weight, and tiller number (Mean values over three varieties of rye), plotted against *B*-chromosome class. (From Muntzing, 1963).

Somatic Cells

An indication of the means whereby *B*'s bring about their effects upon growth and development is provided by comparisons of nuclear components between rye plants with varying *B* frequency (Kirk and Jones, 1970). In Fig. 5 are the average quantities of total protein and of *RNA* in nuclei of plants with different *B* frequencies. It is reasonable to assume that protein and *RNA* amounts are reflections of the rate of genetic activity within the nuclei and it will be observed that, in general, the total protein and *RNA* decrease with increasing *B* frequency. The decrease is, however, by no means regular. It is disproportionately severe in plants with odd numbers of *B*-chromosomes. In this respect the variation within nuclei corresponds

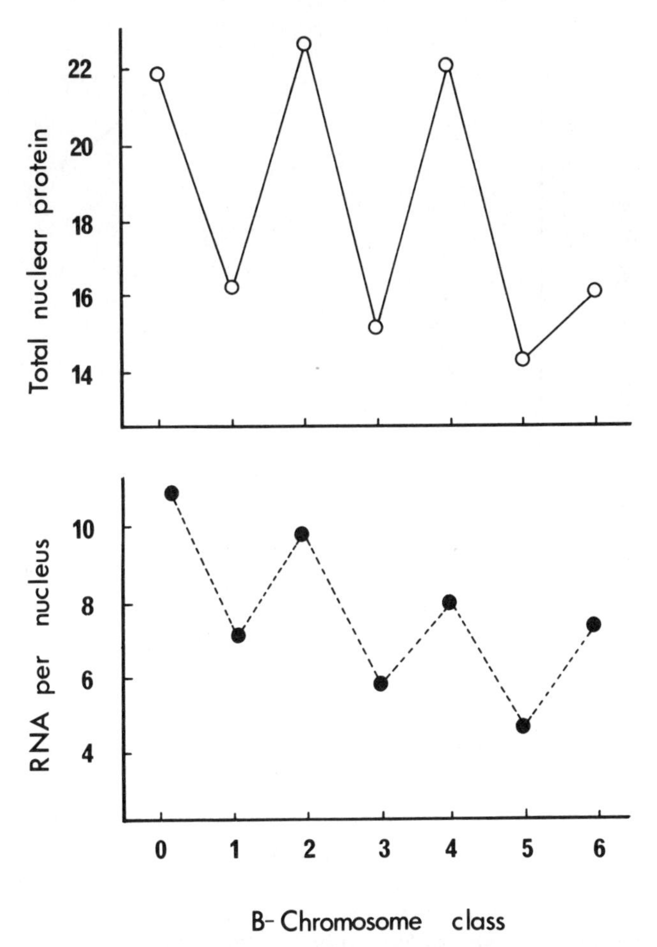

Fig. 5. Mean total nuclear protein ○—○ and nuclear *RNA* amounts ●--● in root tips of rye plants with different numbers of *B*-chromosomes. Protein and *RNA* in arbitrary units.

closely with the variation exhibited by the whole plant in response to odd and even numbers. In Fig. 6 are plotted the histone contents of the *B* plants. The changes in nuclear histone are almost a direct contrast to those in total protein and *RNA*. The histone is *increased* with increasing *B* frequency and disproportionately *high* in plants with *B*'s in odd numbers. It is claimed that histones in the nuclei of higher organisms may regulate the expression of genes in higher organisms by suppressing

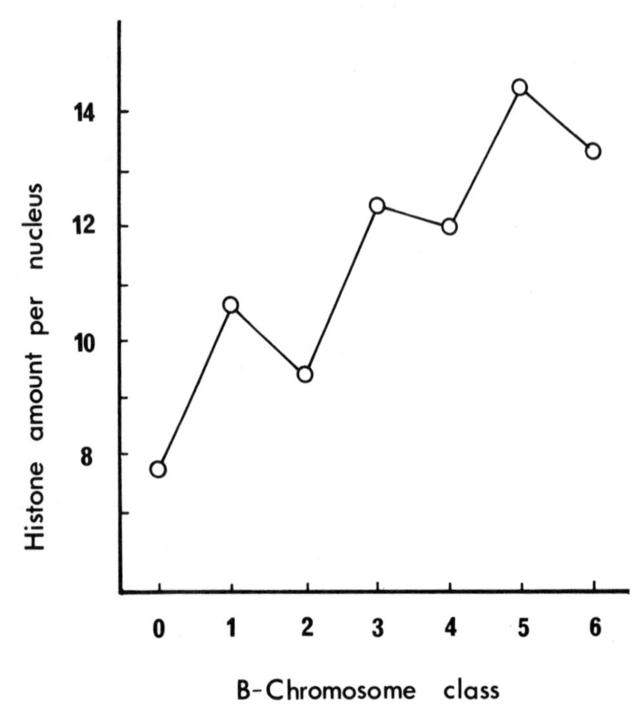

Fig. 6. Histone content per nucleus in root tips of rye (in arbitrary units) plotted against *B*-chromosome frequency.

the transcription of information in chromosomal *DNA* (Bonner *et al.*, 1968; Gilmour, 1969). The pattern of histone variation among the *B*-chromosome classes is compatible with this view. Comparable results have been derived in Maize (Ayonoadu, unpublished).

Recombination

In terms of physiology, of growth and of fertility it is clear from the summary above that the effects of *B*'s overall are deleterious. The question arises therefore, why it is that they are so widely distributed. They must, we suppose, confer some special adaptive benefit upon certain populations. Darlington (1956) was the first to suggest that *B*'s may in fact serve to regulate in an adaptive way the degree of heritable variation. Support for this view comes from the work of Moss in rye (1966). Moss showed that the variability within progenies derived from parents carrying *B*'s was greater than those derived from parents without *B*'s. This increased variability was independent of that attributable to the *B*'s within the progenies themselves. The mechanism which accounts for this boosting of heritable variation in plants has been established by Jones and Rees in rye (1967), by Ayonoadu and Rees (1968) in maize (see also, Hanson, 1962). It had earlier been

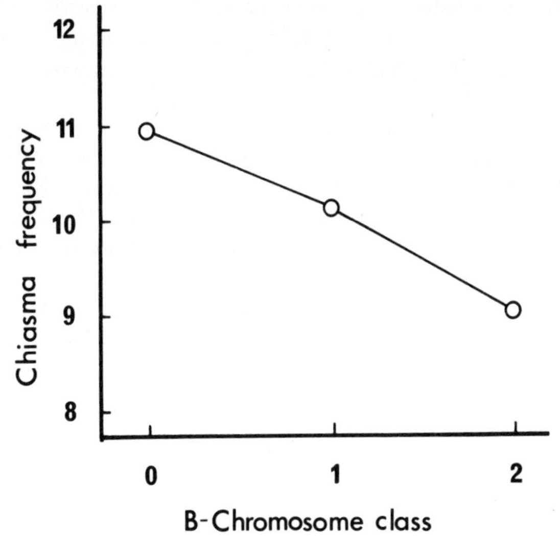

Fig. 7. The average chiasma frequencies in pollen mother cells of *Lolium perenne* plants with varying numbers of *B*-chromosomes.

established in animals, in grasshoppers, by John and Hewitt (1965). In brief, *B*-chromosomes increase the amount of recombination at meiosis in some or all of the bivalents of the normal *A* complement and, thereby, increase the rate of release of heritable variation. Cameron and Rees (1967) have presented results on *Lolium* which, on the face of it, contradict the work in rye and maize. In *Lolium B*'s cause a reduction in recombination at meiosis (Fig. 7). But in *Lolium,* as in the other species, there is nevertheless a regulatory function attributable to *B*'s albeit a restriction as distinct from an increase in genetic recombinatioon.

Chromosome Amplifications

It is not uncommon to find chromosomes split transversely across the centromere to produce one-armed telocentrics. By suitable breeding the telocentrics produced from *B*-chromosomes or from chromosomes of the normal complement, may be added to the standard complement and, in this way, are useful for certain experimental purposes (Riley and Chapman, 1966). In other than experimental populations gain of whole chromosome arms would appear to be rare, even negligible.

The gain or amplification of smaller segments within the chromosomes themselves is another matter. It is evidently widespread among the flowering plants. Along with polyploidy it accounts undoubtedly for the greater part of the *DNA* variation between Angiosperm species.

Between species in higher plants there is, of course, an enormous variation in chromosome size and Sunderland and McLeish (1961) were among the first to show that this size variation was attributable to differences in the *DNA* content of the chromosomes. The nuclear *DNA* content of *Lilium longiflorum* which has large chromosomes, is 57 times greater than that in *Lupinus albus* which has very small chromosomes. This kind of *DNA* variation is found even between very closely related species within a genus. Plate 1 shows the chromosome

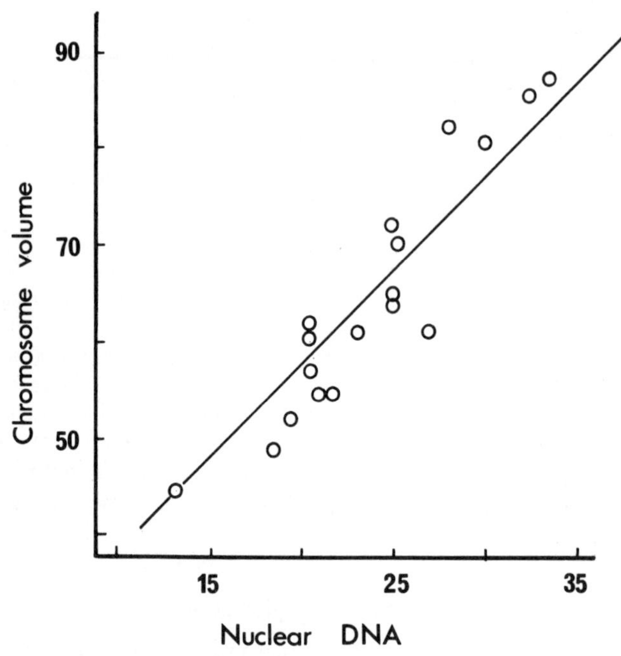

Fig. 8. The mean nuclear *DNA* content (in arbitrary units) plotted against the total chromosome volume in eighteen *Lathyrus* species.

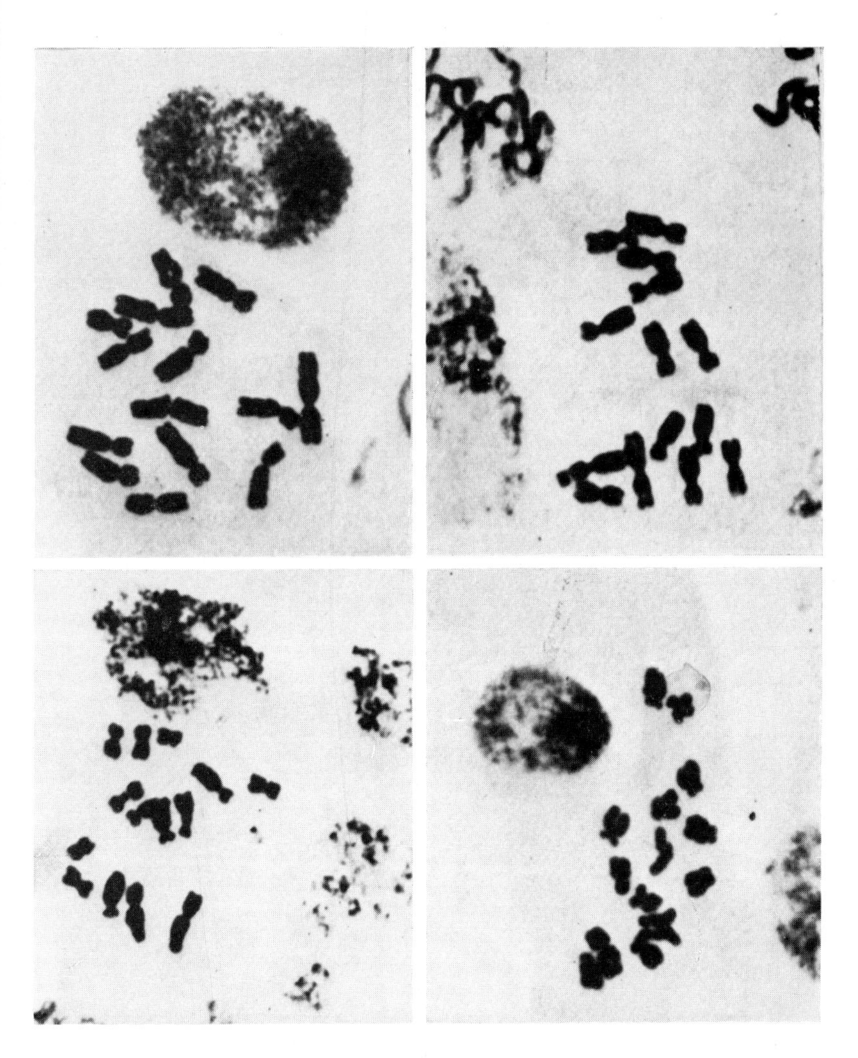

PLATE 1. Metaphases in root tip cells of (1) *Lathyrus hirsutus*, (2) *L. ringitanus*, (3) *L. articulatus*, (4) *L. angulatus*. × ca. 2600]

[*To face p. 200*

complements of four diploid species of the genus *Lathyrus* (2N = 14). The chromosomes of *Lathyrus hirsutus* are three times larger by volume than those of *Lathyrus angulatus.* As Fig. 8 shows this size variation reflects directly a variation in nuclear or chromosomal *DNA*. A comparable *DNA* variation has been described in the genus *Allium* (Jones and Rees, 1968). Figure 9 shows that a twofold *DNA* variation exists between the diploids in this genus. Indeed some of the diploid species have more nuclear *DNA* than some of the tetraploids.

The conclusion that this *DNA* variation is caused by the lengthwise repetition or amplification of chromosome segments is established from

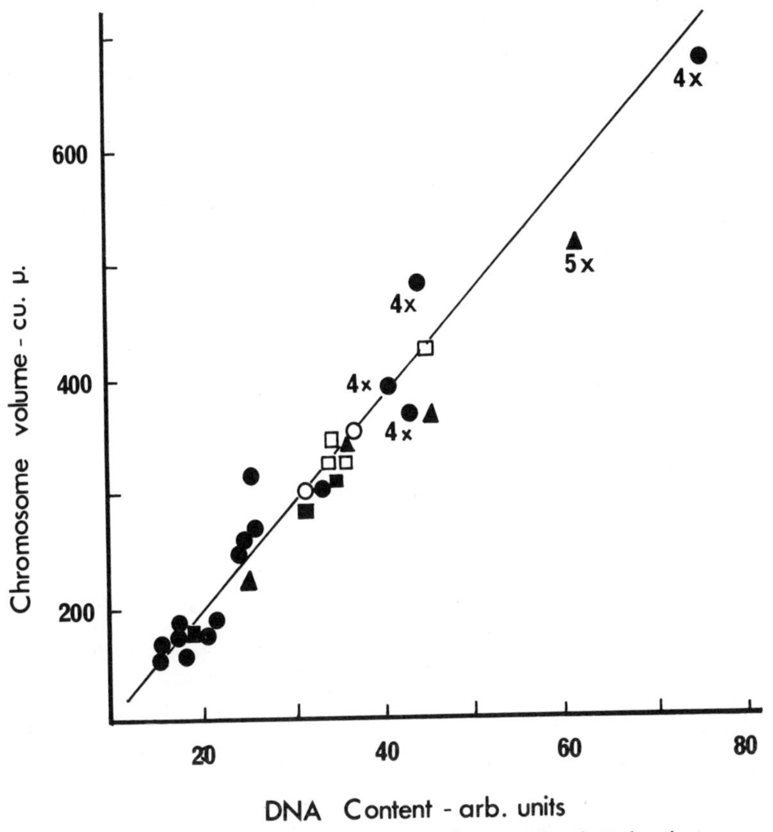

Fig. 9. Mean chromosome volumes of *Allium* species plotted against their *DNA* values. Diploids unless indicated otherwise on the graph. ■ X = 7; ● X = 8; ▲ X = 9; □, ○ are species for which only the chromosome volume, or only the *DNA* values are known.

investigations of F_1 hybrids between species with high *DNA* amounts and large chromosomes on the one hand with, on the other hand, species having low *DNA* and smaller chromosomes (see Rees and Jones, 1967; Jones and Rees, 1968). The basis for this conclusion is as follows. At pachytene of meiosis in the species hybrids the repeated or amplified segments in the large chromosomes are manifested as loops

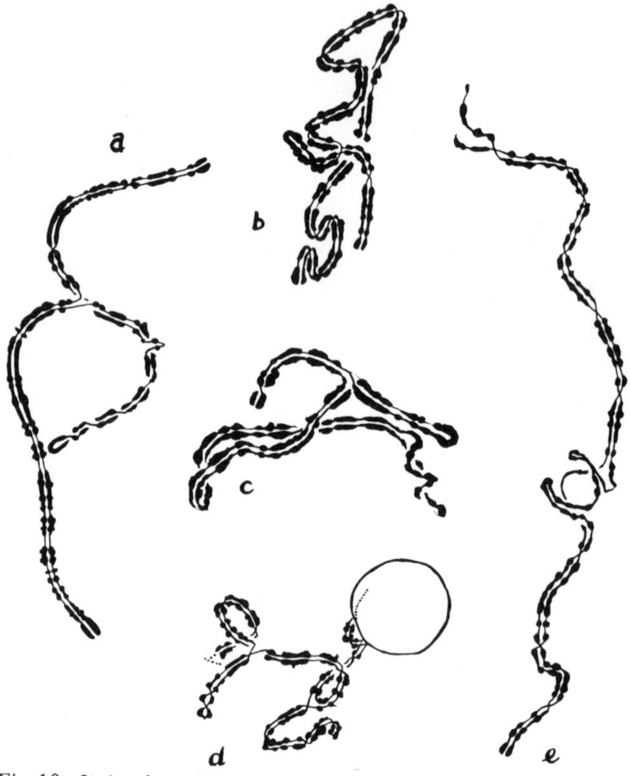

Fig. 10. Isolated pachytene bivalents of *A cepa* x *A. fistulosum.* Note large 'duplication' loops and terminal overlaps.

and overlaps. In some cases these loops are 60 to 70% of the total length of the paired homologous chromosomes (Fig. 10). This figure shows that the amplification appears to be localised, highly restricted to a few segments or loci within the chromosomes, suggesting either that some loci are particularly prone to repetition or else that such amplification is tolerable at some loci and not at others.

While the repetition is highly localised within chromosomes it appears to be widespread between chromosomes of the complement, because all pachytene configurations in the *Allium* hybrids display the characteristic loops and overlaps.

The amplification of chromosome material is now, of course, well established among animal species (see Britten and Kohne, 1969). Detailed work on the mouse (Jones, 1970) and *Chironomus* (Keyl, 1965) provides further unquestionable evidence for its restriction to localised regions within the chromosomes of those species. While commonplace in animals also, the phenomenon gives rise to much smaller variation in chromosomal *DNA* than is the case in higher plants. As with numerical change the higher plants are far more tolerant of quantitative *DNA* change within chromosomes than are animals, especially the higher animals.

We should point out in passing that apart from gain of chromosome material by localised lengthwise repetition there are well established cases of loss of segments accompanying the divergence and evolution of species, e.g. in *Lathyrus* (Rees and Hazarika, 1969).

We should like to emphasise also that we are dealing with a cause of widespread and extensive change in chromosome material. To date, however, it is by no means clear what this kind of change signifies from the genetical point of view. We can point to one or two of the consequences.

First, loss or gain of chromosomal *DNA* has an inevitable effect upon the mitotic cyle, whose duration is positively correlated with *DNA* amount (Van't Hof, 1965; Ayonoadu and Rees, 1967). Second, in certain cases this variable component within the chromosomes is of a heterochromatic nature, e.g. in *Trillium* (Haga, 1967). Heterochromatin is known to exert a regulatory function over genetic activity within the nucleus (Lyon, 1968), and it is not unreasonable to attribute a regulatory role to at least some of the chromosome material acquired by segmental repetition. Thirdly, the repetition may involve structural genes with a consequent direct amplification of their products.

Conclusions

Gain in chromosome material entails a repetition of genetic information or the introduction of new information, and the consequence of gain will clearly depend on the kind of information repeated or introduced. *B*-chromosomes, for example, are relatively inert from which we may infer either that they carry little information or else that the information they bear is largely unutilised. In contrast, the repetition of even a single locus may have startling effects. The Bar eye duplication in *Drosophila* is a convenient example. We must also face the possibility that some of the chromosome material in the nucleus is completely uninformative, genetically meaningless, but this, on any conventional view of adaptation and evolution, is difficult to accept.

In higher plants, as we have seen, gain of genetic material is tolerated to an exceptional degree. It is worth speculating briefly why the higher animals, the mammals, should display an equally remarkable intolerance. To take one source of chromosome gain, polyploidy, its absence in animals is usually explained away on two grounds (see White, 1954). First, all mammals comprise unisexual individuals. Consequently where a rare tetraploid might arise it is committed to mating with a diploid. The resulting triploid offspring will be sterile and the polyploidy would vanish abruptly. Second, polyploidy would upset the differentiation of the sexes where the mechanism depends on a balance between autosomes and *X*-chromosomes. In mammals, however, it is now clear that sex in the main depends on the presence or absence of a *Y*-chromosome so that this second objection is no longer valid (see Darlington, 1953). Indeed we may go further and state that neither of these explanations is sufficient to explain the absence of polyploidy in mammalian populations. In man (Schindler and Mikamo, 1970) and, also, in mouse (Fischberg and Beatty, 1952) it is now established that numerous polyploid embryos are inplanted. We know of no instance where polyploid mouse embryos have reached full term. There are only four instances of human polyploids (triploids) that have come near to a full gestation. All four were highly abnormal, all four died. It is clear therefore that in mammals the polyploidy is incompatible with normal development. We suggest two possible explanations.

(a) There is incompatibility between the diploid mother and the polyploid embryo. In this context it is of interest to note that in the recently reported case of a triploid infant the placenta was abnormally enlarged (Schindler and Mikamo, *loc. cit.*).

(b) The change in cell size consequent upon polyploidy may render incompetent some of the highly specialised cells and tissues of mammalian organisms. In higher plants, of course, there is nowhere near the same variety or complexity of cell organisation as in animals.

Finally it is worth pointing out that although polyploidy, aneuploidy and the acquisition of *B*-chromosomes are of little or no consequence in mammals there is, as already mentioned, good evidence to show that gain of chromosome material by amplification of localised segments is prevalent in at least some mammals. In the mouse, for example, it is estimated that about 10% of the chromosome material is of a repetitious nature (Walker *et al.*, 1969). Exactly what role it fulfils is not, as yet, established.

REFERENCES

Ahloowalia, B. S. (1970). *Abstract, Genetical Society, March 1970.*
Ayonoadu, U. and Rees, H. (1967). *Expl Cell Res.* **52,** 284.
Ayonoadu, U. and Rees, H. (1968). *Genetica,* **39,** 75.
Barton, D. W. (1950). *American Journal of Botany,* **37,** 639.
Battaglia, E. (1964). *Caryologia,* **17,** 245.
Bonner, J., Dahmus, M. E., Fambrough, D., Huang, R. C., Marushige, K. and Tuan, D. Y. H. (1968). *Science* **159,** 47.
Britten, R. J. and Kohne, D. E. (1969). In 'Handbook of Molecular Cytology' (ed. Lima de Faria). North Holland Publishing Company.
Cameron, F. M. and Rees, H. (1967). *Heredity* **22,** 446.
Crowley, J. G. and Rees, H. (1968). *Chromosoma* **24,** 300.
Darlington, C. D. (1953). *Nature* **171,** 191.
Darlington, C. D. (1956). 'Chromosome Botany', Allen and Unwin.
Darlington, C. D., Hair, J. B. and Hurcombe, R. (1951). *Heredity* **5,** 233.
Darlington, C. D. and Mather, K. (1949). 'The Elements of Genetics', Allen and Unwin.
Fischberg, M. and Beatty, R. A. (1952). *Evolution* **6,** 316.
Gilmour, S. (1969). *New Scientist, May,* **15,** 346.
Haga, T. (1967). *Chromosomes Today* **2,** 207.
Hakansson, A. (1948). *Hereditas* **34,** 35.
Hanson, G. P. (1962). *Maize Genetics Co-operation Newsletter* **36,** 34.
Hasegawa, N. (1934). *Cytologia* **6,** 68.
Hazarika, M. H. and Rees, H. (1967). *Heredity* **22,** 317.
Janaki Ammal, E. K. (1950). *The Rhododendron Year Book, 1950,* 92-98.
John, B. and Hewitt, G. M. (1965). *Chromosoma* **17,** 121.
Jones, K. W. (1970). *Nature* **225,** 912.
Jones, R. N. and Rees, H. (1967). *Heredity* **22,** 333.
Jones, R. N. and Rees, H. (1968). *Heredity* **23,** 591.
Jones, R. N. and Rees, H. (1969). *Heredity* **24,** 265.
Keyl, H. G. (1965). *Experientia* **21,** 191.
Kirk, D. and Jones, R. N. (1970). *Chromosoma* **31,** 241.
Kishikawa, H. (1965). *Agricultural Bulletin of Saga University, Oct. 1965,* 1.
Lewis, K. R. (1967). *The Nucleus* **10,** 99.
Lyon, M. F. (1968). *A. Rev. Genet.* **2,** 31.
Moss, J. P. (1966). *Chromosomes Today* **1,** 15.
Muntzing, A. (1946). *Hereditas* **32,** 97.
Muntzing, A. (1957). *Hereditas* **43,** 682.
Muntzing, A. (1963). *Hereditas* **49,** 371.
Norrington-Davies, J. and Crowley, J. G. (1969). *Irish Journal of Agricultural Research* **8,** 359.
Rees, H. and Hazarika, M. H. (1969). *Chromosomes Today* **2,** 158.
Rees, H. and Jones, R. N. (1967). *Nature* **216,** 825.
Rick, C. M. and Barton, D. W. (1954). *Genetics* **39,** 640.
Riley, R. and Chapman, V. (1966). In 'Chromosome Manipulations and Plant Genetics' (eds Riley and Lewis). Oliver and Boyd.
Schindler, A. M. and Mikamo, K. (1970). *Cytogenetics* **9,** 116.
Stebbins, G. L. (1963). Columbia University Press.
Stebbins, G. L. (1966). *Science* **152,** 1463.
Sunderland, N. and McLeish, J. (1961). *Expl Cell Res.* **24,** 541.
Swanson, C. P. (1960). 'Cytology and Cytogenetics'. Macmillan.
Van't Hof, J. (1965). *Experimental Cell Research* **39,** 48.

Walker, P. M. B., Flamm, W. G. and McLaren, A. (1969). In 'Handbook of Molecular Cytology' (ed. Lima de Faria, 1969). North Holland Publishing Company.
White, M. J. D. (1954). 'Animal Cytology and Evolution', 2nd. Edition. Cambridge: University Press.

DISCUSSION

Milne: Is there any reason why chromosomes have been studied more intensively in monocotyledons?
Rees: The only reason is that monocotyledons tend to have larger chromosomes.
Muir: Would you expect a tetraploid human fetus to survive after removal from the maternal invironment?
Rees: I really cannot answer that question in terms of humans. However, it may be relevant that in plants if you cross a maternal tetraploid with a paternal diploid there is a very good chance of getting a lot of seeds but if you cross a diploid female with a tetraploid male in most cases you get no offspring whatever. In the latter case, however, if you remove the offspring from the female and allow them to develop in culture, growth is perfectly normal, so it appears that the adverse effect is due to incompatibility with the maternal diploid environment.
Yielding: Perhaps we ought to study oviparous mammals such as the duck-billed platypus in which embryonic development is less dependent on maternal influence.
Muir: Is the nucleotide composition of *DNA* in the large chromosomes with loops different from that of other chromosomes?
Rees: We have not examined this. The nucleotide composition of *B* chromosomes has been studied, however. There is good evidence that a substantial proportion of their *DNA* consists of reiterated sequences, as in mouse satellite DNA.
Polani: Which chromosomes develop loops?
Rees: We have not been able to determine this. In our material it was not possible to identify individual pairs. However, we hope to be able to do this in future studies.
Allison: It seems that duplication was localised rather than occurring generally.
Rees: Yes, this is so. It reminds one of the very discrete localisation of mouse satellite DNA around the centromeres.
Allison: I should like to go back to the question of mitotic timing in trisomic cells. Being unable to make any sense of the biochemical data

which we have discussed in subjects with Down's syndrome I would like to ask whether there are any data which would be incompatible with the concept that G-trisomic cells have a built-in limit to the speeds with which they can divide, so that if there are any events during embryonic development which require an increased rate of cell division, they may not be able to keep up.

Polani: I can think of no data which would be incompatible with this concept. Indeed, I think it would fit in very well with many observations.

Perhaps I can introduce some other data at this point. Mittwoch has suggested that the sex-determining effect of the *Y*-chromosome in mammals is not due to the expression of structural or regulatory genes on that chromosome, but to the behaviour of the *Y*chromosome as a heterochromatic chromosome. Her hypothesis (which she has tested experimentally) (Mittwoch, 1970, *Phil. Trans. B.* **259**, 113) is that the function of the *Y* chromosome is to speed up the formation of the male gonad by causing acceleration in the processs of cell division and cell growth of the gonad region. It should be noted that in normal mammals the formation of the male gonad occurs at an earlier developmental stage than that of the female, and that in the abnormal situation where there is a single *X*-chromosome, as in the case of human subjects with an *XO* chromosome complement, the gonads are female. The second piece of evidence I should like to quote is a study by Barlow (1970, personal communication), which although still in its preliminary stages seems to indicate that *XO* cells in culture divide more rapidly, that is have a shorter cell cycle, than *XX* cells. He has not yet studied the mitotic times of cells with an abnormally increased number of *X* chromosomes, e.g. trisomy *X*. One could argue that these are likely to have an increase in the mitotic time. These points, which I must emphasise have only indirect bearing to the question of development in G-trisomy, suggest the possibility that the presence of an additional chromosome may slow down the cell cycle and produce developmental abnormalities. (Polani, 1968, *Guy's Hosp. Rep.* **117**, 323).

Cavanagh: Of what order of time is the difference between the cell cycle in *XO* as compared with *XX* cells? Is it about 10% or more than this?

Polani: Barlow's work is still in its early stages. I would not like to be dogmatic about this point but it seems likely that the differences are more than 10%.

Cavanagh: One can consider development to proceed by a series of steps. If there is interference with these steps early in the course of brain development it is likely that there would be far greater disturbance than if interference were to occur later on.

Allison: Possibly a delay in cell-division time in G-trisomy cells could account for the reduced activity of *DNA*-polymerase in Down's syndrome lymphocytes recently described by Agarwal *et al.* (1970, *J. Clin. Invest.* **49**, 161).

Muir: Would you expect teratogenic substances to have a greater effect when the mitotic cycle is shortened? If so, would the tissues of males be more susceptible to teratogenesis than female tissues?

Polani: This is an interesting suggestion. I don't know of any experimental evidence on this point.

Yielding: A similar question was asked in bacteria, namely whether mutation rates were higher when cell division was accelerated but experimental evidence indicated that this was not the case (Novick and Szilard, 1950, *Proc. Nat. Acad. Sci. US.* **36**,708).

Allison: The effect of increased generation time might not be manifest as a specific abnormality. It might affect the timing of some morphological process as we discussed earlier.

Cavanagh: Do other types of chromosomal anomalies have similar kinds of brain disturbances to those found in G-trisomy and if so can anything be learnt from this?

Polani: There is considerable speculation about this point but very little definite data and I don't think it is profitable to draw any conclusion at the present state of knowledge.

General Discussion

Jepson: I think the first two sessions on biochemical studies at the level of tissue slices and sub-cellular preparations were very appropriate for the beginning of the Study Group. They were referred to frequently during the latter sessions and enabled many points to be made which related fundamental biochemical observations to phenomena discussed by the Study Group at other levels.

Dr. Bradford, you will remember, made the case that synaptosomes were intermediate between genuine organelles and artefacts produced during disruption. I think, however, that they may be considered to be no greater artefacts than, for example, microsomes. He showed that isolated synaptosomes had many of the metabolic characteristics that are present in brain slices; for example response to electrical stimulation. These preparations struck me as having great potential as an experimental tool which could be applied to some of the situations we have been discussing. Possibly the reason why this was not suggested by the members of the Study Group may have been that synaptosomes are relatively new preparations and have not yet been considered effectively for experimental investigation in human abnormalities. The synaptosomes from brains of subjects with Hartnup's disease or from animals which are nicotinamide deficient might be useful for testing whether nicotinamide would correct any abnormalities which might be present. Synaptosomes might also be useful to study the suggestion made by Professor McIlwain that the function of gangliosides might be to influence the ionic environment.

The next speaker, Professor McIlwain, described his studies on amino acid transport and protein breakdown induced by electrical stimulation. Judging by amino acid incorporation it seems as if protein synthesis was arrested, but we later discussed the fact that a reduction in precursor incorporation could also be achieved by an increase in protein breakdown. This made me wonder whether electrical stimulation might produce some changes in the lysosomes, for example in their stickiness, in a way similar to that described by Dr. Allison. This might encourage interaction with denatured lysosomal membrane protein and therefore produce an acceleration in protein breakdown by lysosomal proteases.

Allison/Dingle: I don't think that the effect of electrical stimulation on lysosomal properties has been investigated.

Allison: One should remember that the main change produced by electrical stimulation is depolarisation of the cell membranes, and I think it is reasonable to speculate that this might have an effect on the attachment of cell organelles, such as lysosomes, to the cell membrane, but I would think it is less likely that electrical stimulation would produce a change of organelles in the centre of the cell.

Dingle: I am doubtful about the value which experiments on the effects of electrical stimulation of isolated lysosomes would have, since conditions would be very different to those in living cells. I entirely agree with Allison that the major effect of electrical stimulation is that of causing depolarisation of cell membranes.

Scriver: To my mind, two points of great value have been emphasised by this Study Group firstly concerning data which have been presented regarding different types of membranes; that is lysosomal, plasma and others. It appears that different types of cell membranes will be found to be structurally and functionally very different to each other.

The second interesting point I noted, was the clear distinction which the Study Group has repeatedly made between influx and efflux; each is a specific function of cell membranes. At one time it was thought that cell permeability in and out of the cell could be considered together, but we have repeatedly emphasised that influx and efflux are distinct functions and may be subjected to separate genetic control.

Allison: Is there any evidence that nerve cells can have membrane transport defects and could this be a mechanism for the production of brain damage in Hartnup's syndrome?

Scriver: This of course is a very important question and I know of no evidence relating to it at the moment.

Allison: The lysosomal response to nerve section has been studied experimentally. Retrograde degeneration of the nerve axon is associated with formation of numerous multivesicular bodies and a reduction in the number of normal-looking lysosomes. There is breakdown of proteins followed by a sudden increase in protein synthesis.

I should like to draw attention to an apparent paradox which has arisen by use of synthetic substrates in a disease probably due to a defect of a lysosomal enzyme. We can consider Pompe's disease as a straightforward example of a specific deficiency of a lysosomal enzyme, namely, α-glycosidase. This results in accumulation of its substrate, glycogen, inside the lysosomes. In metachromatic leucodystrophy a deficiency has been demonstrated in the activity of the enzyme leucocyte arylsulphatase; an enzyme which has been thought to have no effect on the desulphation of the storage substances which accumulate in this disease. However, it appears that a cerebral sulphatase is lacking in metachromatic leucodystrophy (Jatzkewitz and Mehl, 1969, *J. Neurochem.* **16**, 19). I find the present state of information regarding the Hunter-Hurler syndrome very confusing. The correcting factor for Hurler cells can be extracted from either normal or Hunter cells and that for Hunter cells from either normal or Hurler cells. The nature of this factor is not known but theoretically one can make a number of generalisations. First, as Dr. Muir has told us it is a thermolabile macromolecule with a molecular weight of about 65,000 and therefore may be a protein. Theoretically it may be a repressor but I don't know of any repressor which can be isolated from a group of cells and made to function in another group of cells. Secondly, it might be an enzyme. Again, although metabolic co-operation has been demonstrated between cells lying adjacent to each other in the same culture it would be a unique circumstance in which the enzyme could be extracted and still be active after it had been added to, and taken up by another group of cells.

Muir: The factor is not inactivated by dialysis so that small molecules are not needed for its function.

Allison: Yes, this brings one to the next possibility—that the correcting factor may produce small molecules which can pass into, and exert their effect on the accepting cells.

Dingle: I wonder if we are not reading too much into the complexity of this corrective factor. Could it not be some quite simple molecule such as a metabolite.

Yielding: Yes. The system seems to be very similar to cross feeding in bacteria.

Allison: We must remember, as I said before, that the factor has a molecular weight of about 65,000 and is thermolabile and is therefore probably not just a simple metabolite.

Muir: One thing which appears inconsistent is that although living Hurler fibroblasts can digest labelled dermatan sulphate prepared from Hurler cells when this was added to the external medium (Dorfman and Matalon, 1969, *Amer. J. Med.* **47**, 69), extracts from disrupted cells were unable to do so (Neufeld, private communication).

Dingle: I don't think this is surprising. The activity of lysosomal enzymes is often too low to be demonstrated at any one time *in vitro*. However, growing cells continue to synthesise enzymes so that in time they can hydrolyse large quantities of substrate.

Muir: It appears to be important to consider the assembly of connective tissue materials for export and also whether connective tissue components are normal in the mucopolysaccharidoses. A recent study of Hurler rib cartilage by electron microscopy (Phillips, unpublished results) shows unusual collagen fibres. This could be the result of the abnormal presence of dermatan sulphate in cartilage that has been reported by Meyer in Hurler's syndrome (Meyer and Hoffman, 1961, *Arthritis and Rheumatism* **4**, 552). Dermatan sulphate-proteins possess a striking ability to induce the immediate formation of native-type fibrils from tropocollagen which chondroitin sulphate-proteins do not have (Toole and Lowther, 1968, *Arch. Biochem.* **128**, 567). If dermatan sulphate were to appear in epiphyseal cartilage early in development, it might thus cause skeletal deformities.

Raine: This is very interesting. I wonder if one could explain the occurrence of skeletal deformities in G_{M1} gangliosidosis (Type 1) where the abnormal keratan sulphate-like substance is present, in contrast to the G_{M2} variety where no such mucopolysaccharide has been identified and where there are no skeletal deformities.

Dingle: Techniques for testing the addition of substances to cartilage in culture are available and one could study the effects of added keratan sulphate. I agree with Dr. Muir that one also has to look at the process of packaging of material for export. It has only recently been possible to isolate the Golgi apparatus and it has been found that molecules transported in Golgi vesicles usually have sugar residues. It will be very interesting to study the function of Golgi apparatus in the types of disorder that we have been discussing.

Yielding: One of the things that interests me is the lack of correlation between elevated enzyme activity and the presence of the trisomic chromosome. I think that it would be very interesting to hear more about enzyme activity and RNA synthesis in trisomies other than G-trisomies.

Polani: One should mention that even if regulatory considerations make quantitative expression of a supernumerary gene difficult to detect one might in certain circumstances be able to detect three different types of alleles in a trisomic individual.

Concerning the specificity of individual chromosomes there are some interesting generalisations that can be made. Surviving infants with Patau's syndrome all seem to be trisomic for chromosome 13 and not for 14 or 15 which are in the same group.

Secondly, in the D/G type of interchange formed in Down's syndrome it appears that the G chromosome is nearly always translocated on to chromosome 14 not 13 or 15.

Finally, the two types of marker chromosomes found in the D group, that is either with prominent or chopped-off short arms, seem to be in general chromosome 15 and not 14 or 13.

Another point I should like to make is that in the interchange type of Down's syndrome there may be deletion of the SAT–the satellite stalk–and since this is where the nucleolar organisers are thought to be located, it would be interesting to compare RNA synthesis in interchange trisomics with that in the primary trisomics with Down's syndrome.

I should now like to comment on three quantitative effects observed in individuals with different numbers of sex chromosomes. The first is the effect on stature:

Height and Numerical Anomalies of Sex Chromosomes, without Mosaicism

Type	*No.*	$\bar{x}$	*S.E.*$\bar{x}$
45,*X*	128	141.80	0.56
46,*XX*		162.20	
47,*XXX*	30	163.07	1.49
46,*XY*		174.70	
47,*XXY*	118	175.69*	0.77
48,*XXYY*	22	180.52	2.12
47,*XYY*	19	182.95	1.67

* 31 institutionalised defectives are included, whose average height was about 172 cm, while for the rest it was about 177 cm.

(From the literature and with personal data (see original publication)) (Polani 1970. In 'Congenital Malformations.' Proceedings of the Third International Conference on Congenital Malformations, The Hague, 7-13 September, 1969. (Eds Clarke Fraser and McKusick), p. 233. Amsterdam and New York: Excerpta Medica Foundation.

As you can see, there is an incremental effect of added sex chromosomes and it looks as if the *Y* chromosomes have the stronger effect.

The second is on a dermatoglyphic character investigated by Penrose who showed that with increasing numbers of sex chromosomes there is a decrease in the 'total ridge count' of the skin of finger-tips. This count, about 160 in *XO* subjects, is depressed by about 30 for each additional *X* chromosome, and by about 20 for each additional *Y* chromosome (Polani, 1969, *Nature* **223**, 680).

The third quantitative effect I should like to comment on is the effect of *X*-chromosome numbers on IgM levels. IgM levels increase in relation to added *X* chromosomes and *XO*'s have similar IgM levels to *XY* individuals (Wood *et al.*, 1970, *Atti Ass. genet. Ital.* **15**, 228; Wood *et al.*, 1969, *Brit. med. J.* **4**, 110; Rhoades *et al.*, 1969, *Brit. med. J.* **3**, 439).

Stern: Can the findings of low IgM in Down's syndrome in any way be tied up with the *X*-chromosome effects on IgM?

Polani: We don't have any idea.

Rees: Should one not approach the question of specific biochemical changes in Down's syndrome by comparing abnormalities in the primary trisomic variety with those in the interchange type and discard any which are not found in both types since they may not be relevant to Down's syndrome?

Polani: This approach has been attempted (Rosner *et al.,* 1965, *New Eng. J. Med.* **273**, 1356) but it has been difficult to confirm many of the conclusions of these workers.

Stern: Sparkes and Baughan, 1969 (*Am. J. Hum. Gen.* **21**, 430) made a very careful attempt to distinguish biochemical differences between two types of translocation Down's syndrome. The type with G21/D from G22/D. They were unable to find any convincing differences but they did have two cases of G21/D with low erythrocyte phosphofructokinase activities which they could not explain.

List of Participants

A. C. Allison, BM, DPhil	Division of Cell Pathology, National Institute for Medical Research, London
P. F. Benson, MB, PhD, MSc, MRCP, DCH	Paediatric Research Unit, Guy's Hospital Medical School, University of London
H. Bradford, BSc, PhD	Biochemistry Department, Imperial College of Science and Technology, University of London
J. B. Cavanagh, MD, MRCP	Director, MRC Research Group in Applied Neurobiology, Institute of Neurology, London
J. T. Dingle, PhD	Strangeways Research Laboratory, University of Cambridge
J. B. Jepson, MA, BSc, DPhil, FRIC	Courtauld Institute of Biochemistry, Middlesex Hospital Medical School, University of London
R. N. Jones, BSc, PhD	Department of Agricultural Botany, University College of Wales, Aberystwyth
G. M. Komrower, TD, MB, ChB, FRCP	Department of Child Health, University of Manchester
H. McIlwain, DSc, PhD	Department of Biochemistry, Institute of Psychiatry, University of London
M. D. Milne, MD, FRCP, MRCS	Department of Medicine, Westminster Medical School, University of London
Helen Muir, PhD	Kennedy Institute of Rheumatology, London

N. B. Myant, DM, BSc, MRCP — Medical Research Council, Cyclotron Building, Hammersmith Hospital, London

P. E. Polani, MD, FRCP, DCH — Prince Philip Professor of Paediatric Research and Geneticist, Paediatric Research Unit, Guy's Hospital Medical School, University of London

R. J. Pollitt, MA, PhD — Medical Research Council Unit for Metabolic Studies in Psychiatry, Sheffield

Ruth Porter, MRCP, DCH — Deputy Director, Ciba Foundation for the Promotion of International Co-operation in Medical and Chemical Research, London

D. N. Raine, MB, PhD, BSc, FRIC — Department of Biochemistry, Children's Hospital, Birmingham

H. Rees, BSc, PhD, DSc — Department of Agricultural Botany, University College of Wales, Aberystwyth

D. Robinson, DSc, PhD — Head of the Biochemistry Department, Queen Elizabeth College, London

C. R. Scriver, BA, MD, CM — Montreal Children's Hospital, McGill University Research Institute, Montreal, Canada

R. G. Spector, MD, MRCP, PhD, MRCPath — Reader in Applied Pharmacology, Guy's Hospital Medical School, University of London

J. Stern, BSc, PhD — Principal Biochemist, Neuropathology Department, Queen Mary's Hospital for Children, Carshalton, Surrey

Y. L. Yielding, MD — Laboratory of Molecular Biology, Medical Center, University of Alabama, Birmingham, U.S.A.

Index

2-Acetamido-1-(β′-L-aspartamido)-1, 2-dideoxy-β-D-glucose (AADG). *See also under Aspartylglycosaminuria*
- degradation of, 96, 97, 100, 101, 102
- in isolated lysosomes, 101

N-Acetylneuraminic acid (NANA), 118, 122, 123

Actinomycin D damaging lysosomes, 69

β-Alaninaemia, 48

Alanine, in brain, 18

Aleutian disease, 108

Alkaline phosphatase in Down's syndrome, 144, 147, 157

Alzheimer's disease, 154
- lysosomes in, 70

Amaurotic family idiocy, terminology, 133

Amino acids,
- absorption in Hartnup disease, 62
- effect of overload on kidney, 28
- effect on membrane potential, 27
- efflux, 65
- inborn errors of transport, 38
- in brain,
  - distribution and utilization, 15-27
  - effect of hypoxia, 17
  - effect of starvation, 17
  - effect of stimulation, 20, 209
  - formation, 19
- membrane transport, 29-42
  - inborn errors of, 38
  - in cystinuria, 38, 65
  - in Hartnup disease, 38
  - in iminoglycinuria, 38
  - in phenylketonuria, 27
  - peptide absorption in, 63
- mental retardation and, 15, 16
- release from synaptosomes, 5-7, 9, 11, 13, 25, 74
- renal clearance in Hartnup disease, 56

Amino acids–*cont.*
- retention by newborn kidney, 35
- tubular reabsorption of,
  - defects in, 43-54
  - iminoglycinuria, 47
  - in cystinuria, 47
  - in Hartnup disease, 46

Amino aciduria, 44

γ-Aminobutyric acid in brain, 12, 18
- in kidney, 12
- release in synaptosomes, 5, 6, 9, 11, 14

Amniotic fluid in diagnosis of sphingolipidoses, 138

Anaesthesia, effect on microtubules, 26, 27

Aneuploidy in plants, 191

Angiokeratoma corporis diffusum, *See Fabry's disease*

Argininaemia, 49

Argininosuccinicaciduria, 28

Aspartate release in synaptosomes, 5, 6, 9, 14

Aspartic acid in brain, 18

Aspartylglycosaminuria, 95-106
- chemical studies, 95-96
- enzyme studies, 96-98
- lysosomal enzyme defect in, 97-100, 102-103
- possible storage aspects, 99-101

Axons, lysosomes in, 73

Bacterial infection in granulomatous disease, 110, 111

Barré syndrome, lysosomes in, 69

Blood,
- enzymes in,
  - in Down's syndrome, 147-153
  - in sphingolipidoses, 136

Blood brain barrier, protein metabolism and, 20

Brain,
- abnormalities in trisomy, 208
- adaptation, amino acids and, 15-27

Brain–*cont.*
amino acids in,
distribution and utilization, 15-27
effect of hypoxia, 17
effect of starvation, 17
effect of stimulation, 20
formation of, 19
protein lability and, 16-20
chlorpromazine in cells, 116
enzymes in Down's syndrome, 154
GABA activity in, 12
gangliosides in, 120
breakdown, 122-123, 124
Gaucher's disease and, 127
synthesis, 121, 122
glutamate in, 23
iminoglycine transport in, 32
in Down's syndrome, 153-155
in Gaucher's disease, 127, 128
in trisomy, 171-172
lysosomes in, 68
neurotransmission, agents involved in, 22
protein synthesis in, 20, 24
transmitter release from terminals, 21

Calcium uptake in synaptosomes, 5
Cataracts, 53
Cathepsin,
in lysosomes, 70
in mucopolysaccharidoses, 79
Cells, ageing of, in Down's syndrome, 157
Cell membranes, damage from electrodes, 1, 25
Ceramide, 133
formation, 120
in gangliosidosis, 125
residues in sphingolipids, 117
structure, 118
Ceramide trihexoside, 119, 126
Cerebellar ataxia in Hartnup disease, 46
Cerebral adaptive processes, protein synthesis and, 16
Chediak-Higashi syndrome, 107-109
Chlorpromazine,
effect on synaptosomes, 8, 11
in brain cells, 116
Cholesterol levels in Down's syndrome, 144
Chondroitin sulphate, 79, 81
Chromosomes,
DNA content, 200
gain in plants, 158, 185-208
height and, 213
in Down's syndrome, 143, 147, 161, 214
in plants,
amplifications, 199-203
gain, 158, 185-208
nucleolar organisers, 168, 169-170
nucleotides in, 206
redundancy
effect on gene expression, 173-184
RNA synthesis in, 170
sex, 213
size of in plants, 200
B-Chromosomes,
distribution and inheritance of, 194
effect on growth and fertility, 195
in plants, 193-199
recombinations, 198
Y-Chromosomes, 207
Citrullinuria, 49
Cocaine, effect on synaptosomes, 8, 11
Coeliac syndrome in cystinuria, 38
'Corrective substances', 87, 89, 90
Cortex, lysosomes in, 13
Cyclic AMP in protein synthesis, 22, 24
Cystic fibrosis, Down's syndrome and, 151-152
Cystinosis, 104
Cystinuria, 52, 65
amino acid transport in, 38
cerebral phenotype, 39-40
coeliac disease in, 38
in mental illness, 39
in schizophrenia, 39
mental retardation and, 52, 53

Demyelinating disease, lysosomes in, 69
Dendrites, thickening, 26
Desoxyribonucleic acid,
content of chromosomes, 200
polymerase,
activity in Down's syndrome, 146, 150, 208
synthesis in trisomy, 157
variation in, 186, 200, 201

Dermatan sulphate, 76, 79, 80, 81, 83, 87, 89, 211, 212
Diabetic neuropathy, lysosomes in, 69
Dipeptides, role in Hartnup disease, 62, 64
Diphtheria toxin damaging lysosomes, 69
Down's syndrome,
- absorption defect, 152
- alkaline phosphatase activity in, 144, 147, 157
- biochemistry of, 143-159, 214
- brain enzymes in, 154
- brain in, 153-155
- cell ageing in, 157
- chromosomes in, 143, 147, 161, 214
- cystic fibrosis and, 151-152
- DNA polymerase activity in, 146, 150, 208
- electroencephalography in, 172
- endocrinological aspects, 153
- enzyme deficiency in, 147
- glucose metabolism in, 171
- intellectual impairment at birth, 159
- protein losing enteropathy in, 152
- RNA synthesis in leucocytes, 161-172
- serotonin pathway in, 147, 154, 159
- serum proteins in, 144-146
- tryptophan metabolism in, 146-147
- uric acid excretion in, 150, 158
- vitamin B_6 in, 146-147

Electroencephalography in Down's syndrome, 172
Enzymes,
- activity,
  - control of, 175-176
  - in gene control, 182-183
  - in trisomy, 182
- catalytic subunits, 94
- deficiencies,
  - in Down's syndrome, 147
  - in inborn errors of metabolism, 131
  - in metachromatic leucodystrophy, 210
  - in sphingolipidoses, 135, 136, 141
  - in Tay-Sachs disease, 139

Enzymes–*cont.*
  - in trisomy, 212
- for ganglioside metabolism, 121, 122, 129
- in aspartylglycosaminuria, 97, 102-103
- in blood in sphingolipidoses, 136
- in brain, in Down's syndrome, 154
- in lysosomes, 68
  - biochemistry, 68
  - immunoenzymology of, 70-72
  - in neural pathology, 69-70
  - defects in, 76, 107
  - in mucopolysaccharidoses, 103
- membrane transport and, 53
- proteolytic, in brain, 18
- urinary excretion in sphingolipidoses, 136

Erythrocytes,
- cations in Down's syndrome, 151
- enzyme deficiency in Down's syndrome, 148

Fabry's disease, 135, 137
- gangliosides in, 124, 125
- glycosaminoglycans in, 83
- lysosomal hydrolase deficiency in, 77
- visceral involvement, 140

Fanconi syndrome, 45
Finger prints, 213
Fluorocitrate damaging lysosomes, 69

α-Galactosidase, 126, 137
β-Galactosidase,
- deficiency of, 77, 80, 125-6, 137, 140

Galactosaemia, 45
Galactose 1-phosphate, cataracts and, 53
Galactosidase,
- deficiency of, 77, 80, 125, 140
- identification, 78
- role in glycosaminoglycan metabolism, 90

Galactosyl transferase, 82
Gammaglobulin in Down's syndrome, 144
Gangliosides, 133
- action of lysosomal hydrolases, 127
- biochemistry of, 117-132
- breakdown, 122-123, 124
- codes for denoting, 119
- definition and structure, 117-120

Gangliosides–*cont.*
distribution, 120
function, 120, 131
in brain, 120
in Tay-Sachs disease, 77
metabolism,
enzymes involved, 129
inborn errors of, 117-132
relationships, 134
storage diseases, 123
structural formulae, 119
synthesis, 120-122
trivial names for, 119
Gangliosidosis, 79, 84, 125
diagnosis, 137
GAG in, 140
relation to malabsorption syndrome, 140-141
renal involvement, 141
skeletal deformities in, 212
terminology, 133
visceral involvement, 140
Gaucher's disease, 135
abnormal ganglioside metabolism in, 124, 125, 127
brain in, 127, 128
diagnosis, 137
Gene,
enzyme activity in control, 182-183
expression,
effect of chromosomal redundancy, 173-184
protein activity, 176
RNA and, 180
regulatory,
redundancy, 179-181
structure and redundancy, 177, 183
supernumerary, 212
in microorganisms, 174
repressor-operator model, 174, 178, 180
translational control, 175, 180
Globoside, 119, 127, 128
Globoside storage diseases, 135, 137
Glomerulus, proline transport in, 32
Glucose metabolism in Down's syndrome, 171
α-Glucosidase, 107
Glutamate,
half life, 23
in brain, 23
metabolism, 12, 13, 26
release in synaptosomes, 5, 6, 9, 14, 26
Glutamate dehydrogenase, 12
Glutamic acid in brain, 18
Glycine
effects on non-ketotic hyperglycinaemia, 27
incorporation of in kidney, 37
ontogeny of efflux of, 35, 36
transport across kidney membranes, 30
Glycogen, lysosomal breakdown, 107
Glycogenosis type II, 107
Glycolysis in synaptosomes, 3, 9
Glycosaminoglycans (GAG), 75
abnormal secretion, 83
catabolism, 79
degradation of, 80, 85, 87, 89
in gangliosidoses, 140
in mucopolysaccharidoses, 75-90
synthesis, 90
types of, 75
urinary carriers of, 83
Golgi apparatus, relation to lysosomes, 68, 73, 115
Granulocytes in Down's syndrome, 150
Granulomatous disease.
infection in, 110, 111
lysosomes in, 109
neutrophils in, 110

Haematoside, 119
Haemoglobin, 111
Hartnup disease, 55-66
amino acid transport in, 38
biochemical lesions, 46
discovery of, 55
family pedigrees, 57, 58, 59
membranes in, 210
mental symptoms in, 55, 57, 65
methionininuria in, 64
nicotinamide in, 61, 65
rash in, 55, 61, 66
real amino acid clearance of, 56
role of dipeptides, 62, 64
tryphophan metabolism in, 59, 62, 64, 66
Height, chromosomal abnormalities and, 213
Heparan, 76, 81
metabolism, 84
Heparan sulphate, 80, 84
Heparin, 79, 80, 84
Hepatolenticular degeneration, 45

Hexosaminidase,
deficiency, 126-127, 135
in Tay-Sachs disease, 131
Histamine, role in brain metabolism, 22
Histones in plants, 197
Homocystinuria, treatment, 50
Hunter's syndrome, 211
β-galactosidase deficiency in, 77, 140
genetics of, 83, 94
glycosaminoglycans in, 76, 86, 87, 89
Hurler's syndrome, 75, 211
chondroitin sulphate in, 81
β-galactoside activity in, 77, 140
genetics, 94
glycosaminoglycans in, 76, 86, 89
isoenzymes in, 77
mucopolysaccharides in, 76, 77
Hydroxykynureninuria, 66
4-Hydroxy-L-proline transport across kidney membranes, 30
Hydroxylysinuria, 54
Hydroxyprolinaemia, treatment, 50
Hyperglycinaemia, non-ketotic, 27
Hyperprolinaemia, 48
Hypoglycaemia in maple syrup urine disease, 27
Hypoxia, effect on amino acids in brain, 17

Iminoglycine transport in kidney, 30-32, 33
Imino-acid-glycinuria, 28, 38
renal, 33
sex incidence of, 54
tubular reabsorption, 47
Immunoglobulins,
in Down's syndrome, 145
in trisomies, 146
sex chromosomes and, 213
Inclusion bodies in lysosomes, 111-112
Infantile amaurotic idiocy, isoenzymes in, 77
Infantile metachromatic leucodystrophy, 77
Information storage, role of gangliosides, 120
Intestinal absorptive membranes, iminoglycine transport across, 32
Isoenzymes in Tay-Sachs disease, 77

Jakob-Creutzfeld disease, lysosomes in, 70
Jansky-Bielschowsky disease, 133
Job's syndrome, 111

Kidney,
amino acid retention in newborn, 35
effect of amino acid load on, 28
GABA activity in, 12
glycine transport across membranes, 30
4-hydroxy-L-proline transport across membranes, 30
iminoglycine transport in, 30-32, 33
in gangliosidoses, 141
in Hartnup disease, 56, 57
proline incorporation in, 37
proline transport across membranes, 30
tubular damage, conditions associated with, 44-46
tubular reabsorption of amino acids, 43-54
tubular transport systems, 43-44
Krabbe leucodystrophy, 77, 135
Kufs disease, 133

Lactate dehydrogenase content of synaptosomes, 6, 7
Lactosyl ceramide, 119, 125
Leucine uptake in brain, 17, 18, 19
Leucocytes,
enzyme deficiency,
in Down's syndrome, 148
in sphingolipidoses, 136, 141
RNA synthesis in Down's syndrome, 161-172
Lipids,
abnormal metabolism, 135
in brain in Down's syndrome, 154
in urine, 138
Lipodoses, lysosomes in, 70
Lowe's syndrome, 45
Lymphocytes,
in Chediak-Higashi syndrome, 108
uridine incorporation into RNA, 163
Lysine, deficiency, 52

Lysosomes,
abnormal neurones and, 113
amino acids in, 74
biochemical studies, 69
cathepsin D in, 70
electrical stimulation of, 210
enzymes in,
biochemistry, 68, 73, 74
defects of, 107
deficiencies, 76
immunoenzymology of, 70-72
in aspartylglycosaminuria, 97-100, 102-103
in mucopolysaccharidoses, 76, 103
in pathological conditions, 69-70
role of, 70
following injury, 69
fusion with other vesicles, 105, 106
hydrolases,
action on gangliosides, 127
glycosaminoglycan catabolism by, 79
in axions, 73
in Chediak-Higashi disease, 109
inclusion bodies due to substrate saturation, 111-112
in cortex, 13
in cystinosis, 104
in injured nervous tissue, 69
in malarial parasites, 115
in mucopolysaccharidoses, 106
in nervous tissue, 67-74
in protein synthesis, 112-113
internal pH, 94
membranes, 104, 105, 106
methods of study, 67
microscopy, 67
protein metabolism in, 104
relation to Golgi apparatus, 68, 73, 115
types of abnormality, 107-116
unstable protein incorporation, 112, 115

Malabsorption syndrome, relation to gangliosidoses, 140-141
Malaria, 115-116
Maleic acid, 45
Maple syrup urine disease, hypoglycaemia in, 27
Maroteaux-Lamy syndrome, glycosaminoglycans in, 76
Membrane potentials,
effect of amino acids on, 27
measurement in synaptosomes, 3
metabolic performance and, in synaptosomes, 1-14
Membrane structure, 29-30, 210
Membrane transport,
enzyme systems and, 53
of amino acids, 29-42
specificity, 53
Meningeal carcinomatosis, lysosomal enzymes in, 70
Mental performance, ribonucleic acid and, 15
Mental retardation in starvation, 17
Metachromatic leucodystrophy, 135, 136
abnormal ganglioside metabolism in, 125
diagnosis, 137, 138
enzyme deficiency in, 210
metabolic basis, 136
Methionine, effects of excess phenylalanine on, 16, 17
Methionininuria in Hartnup disease, 64
Mitochondria, respiration in, 13
Morquio's disease,
glycosaminglycans in, 76, 78, 84
Mosaicism in metabolic disease, 170
Mucopolysaccharidoses,
aetiology of, 83, 85, 93-94
biochemistry of, 75-94
corrective factor, 87, 89, 90
definition of, 75
β-galactosidase deficiency in, 77-78
genetic defect, 80-83
glycosaminoglycans in, 75-90
lysosomes in, 103, 106
mental retardation in, 84
Myeloma, Chediak-Higashi disease and, 109

Nerve terminals, in brain, transmitter release from, 21
Nervous tissue, lysosomes in, 67-74
Neurones,
abnormal, lysosomes and, 113
protein requirements, 20
survival of damage to, 1
Neurotransmission, agents involved in, 22

Nicotinamide,
deficiency in Hartnup disease, 47, 61, 65
in Down's syndrome, 146
Niemann-Pick disease, lysosomes in, 70
Noradrenaline, role in brain metabolism, 22
Nucleotides,
composition of RNA, 163, 166
in chromosomes, 206

Ontogeny, as probe of transport specificity, 33-38

Patau's syndrome, 171, 212
Pellagra, 55, 61, 66
Phagocytosis,
granulomatous disease and, 110
Phenylalanine excess of, 16, 17
Phenylketonuria, 27, 50
Plants,
aneuploidy in, 191
B-chromosomes in, 193-199
chromosome gain in, 158, 185-208
chromosome size in, 200
chromosomal nucleolar organisers in, 169-170
chromosome amplifications in, 199-203
histones in, 197
polyploidy in, 188-191
trisomy in, 192
Plasma, enzymes in, in sphingolipidoses, 136
Polyploidy, 204
advantages of, 190
in plants, 188-191
Pompe's disease, 210
Potassium, content of synaptosomes, 4, 5
metabolism in Down's syndrome, 151, 158
Proline,
effect on brain metabolism, 28
excess of, 48
incorporation in kidney, 37
ontogeny of efflux of, 35, 36
transport across kidney membranes, 30
Protein,
activity in gene expression, 176
lability, cerebral amino acids and, 16-20

Protein–*cont.*
stability, control of, 176
synthesis,
by synaptosomes, 13
cerebral adaptive processes and, 16
in brain, 20, 24
in neurons, 20
lysosomes and, 112-113
organisation of, 168
role of cyclic AMP, 22, 24
turnover, lysosomes and, 112-113
unstable, lysosome incorporation, 112, 115

Renal iminoglycinuria, 33
Respiration,
in synaptosomes, 3, 9, 13
neutrophil, in granulomatous disease, 110
Reticulocyte membranes, iminoglycine transport in, 32
Ribonucleic acid,
gene expression and, 180
in trisomy, 171
mental performance and, 15
nucleotide composition, 163, 166
sedimentation pattern, 164
sucrose density gradient analysis, 163
synthesis,
in chromosomes, 170
in Down's syndrome, 161-172
in trisomy, 212
^{3}H-uridine incorporation, 163

Saliva, in Down's syndrome, 151
Sanfilippo syndrome,
β-galactosidase deficiency in, 77
glycosaminoglycans in, 76, 83, 87, 89
Scheie syndrome, glycosaminoglycans in, 76, 87
Schizophrenia, cystinuria in, 39
Serotonin,
pathway in Down's syndrome, 147, 154, 159
role in brain metabolism, 22
Serum proteins in Down's syndrome, 144-146
Sodium metabolism, in Down's syndrome, 151, 158
Sphingolipids, 117

Sphingolipidosos,
age incidence, 129
enzyme deficiency in, 135, 136, 141
laboratory diagnosis, 133-141
analytical, 135
material available, 136
methods, 137
Sphingosine, 117
formation of ceramide from, 120
structure, 118
Spielmeyer-Vogt disease, 133
Starvation, effect on ,amino acids in brain, 17
Substrate metabolism, transport ontogeny and, 35-38
Sucrose density gradient analysis, 163
Synaptosomes, 209
amino acid release in, 5-7, 9, 11, 13, 25
as cell-like entities, 2-3
calcium uptake, 5
effect of temperature on, 7
electrical stimulation to, 3, 7, 11, 13
glutamine metabolism in, 26
glycolysis in, 3, 9
lactate dehydrogenase content, 6, 7
lysosomal enzyme activity in, 13
membrane potentials and metabolic performance in, 1-14
populations of, 26-27
protein synthesis in, 13
resealing of, 25
respiration in, 3, 9, 13
sodium and potassium content, 4, 5

Tay-Sachs disease, 76, 133, 135
abnormal ganglioside metabolism in, 123
age incidence, 129
enzyme deficiency, 139
gangliosides in, 77, 126, 131
hexosaminidases in, 131
lysosomes in, 70
variants, 135
Tay-Sachs ganglioside, 119
Tears, protein in, 151
Temperature, effect on synaptosomes, 7
Tetracycline, degradation products, 45
Tetraploidy, 206
Tetrazolium damaging lysosomes, 69
Tetrodotoxin effect on synaptosomes, 8, 11
Thalassaemia, 111, 112, 115
Thyroid activity in Down's syndrome, 153
Transport mutations, phenotypic effects of, 38-39
Transport systems,
peptide absorption in amino acids, 63
in kidney, 43-44
in single amino acidopathies, 48-49
ontogeny as probe for specificity, 33
saturation in therapy, 50
Trisomy,
brain in, 171-172, 208
cell division in, 207, 208
chromosomal nucleolar organiser site in, 168
DNA synthesis in, 157
enzyme activity in, 182, 212
immunoglobulins in, 146
in plants, 192
RNA in, 171, 212
Tryptamine, Hartnup disease and, 61, 64, 66
Tryptophan, in Hartnup disease, 62
Tryptophan degradation products, in Hartnup disease, 59, 60
Tryptophan metabolism in Down's syndrome, 146-147

Uric acid excretion in Down's syndrome, 150, 158
^{3}H-Uridine, incorporation in RNA, 163
Urine,
enzyme excretion in, in sphingolipidoses, 136
lipids in, 138

Valine uptake in brain, 17, 18, 19
Vitamin B_6 in Down's syndrome, 146-147

Wilson's disease, 45